BIOLOGICAL NITROGEN FIXATION

Ecology, Technology, and Physiology

BIOLOGICAL NITROGEN FIXATION

Ecology, Technology, and Physiology

Edited by

Martin Alexander

Cornell University
Ithaca, New York

PLENUM PRESS·NEW YORK AND LONDON

Library of Congress Cataloging in Publication Data

Main entry under title:

Biological nitrogen fixation.

"Proceedings of a training course. . . held January 18–29, 1982, in Caracas, Venezuela"—T.p. verso.
  Bibliography: p.
  Includes index.
  1. Nitrogen—Fixation—Congresses. 2. Rhizobium—Congresses. 3. Micro-organisms, Nitrogen-fixing—Congresses. 4. Legumes—Congresses. I. Alexander, Martin, 1930-
QR89.7.B54 1984   589.9′504133   84-3378
ISBN 0-306-41632-8

Proceedings of a training course on
Biological Nitrogen Fixation and its Ecological Basis,
held January 18–29, 1982, in Caracas, Venezuela

© 1984 Plenum Press, New York
A Division of Plenum Publishing Corporation
233 Spring Street, New York, N.Y. 10013

Printed in the United States of America

FOREWORD

The Instituto Internacional de Estudios Avanzados, founded by decree of the President of the Republic of Venezuela, Dr. Luis Herrera Campins, on November 15, 1979 and constituted on July 8, 1980, sponsors, within the Educational Program for the Third World Countries, a series of courses on Scientific Basis for Agricultural Productivity, the first of which "BIOLOGICAL NITROGEN FIXATION AND ITS ECOLOGICAL BASIS," was held in Caracas on January 18-29, 1982.

We are pleased to present, in this volume, the proceedings of this course. It is a pleasure to thank each and every one of the persons who collaborated in the realization of this event, especially: Dr. Martin Alexander (Cornell University, U.S.A.); Dr. Luis Bascones (FUSAGRI, Venzuela) and Dr. Ivan Casas (Universidad del Zulia, Venezuela), who assisted me during the preparation and execution of this activity; the teaching staff, whose presentations are the chapters of this volume; and the participants from Argentina, Brazil, Colombia, Costa Rica, Egypt, Ghans, Guatemala, India, Indonesia, Mexico, Nicaragua, Nigeria, Pakistan, Peru, Syria, and Venezuela who shared this course with us. Special recognition is given to UNESCO and FUSAGRI for their valuable contribution.

I sincerely hope this publication will be the first of a series on the same level that will originate from the courses in this same program.

Leopoldo Villegas

PREFACE

One of the major nutrients needed for the production of food and feed in most of the world is nitrogen. Indeed, more often than any other element, nitrogen is the nutrient element that limits the quantity and quality of food and feed that can be produced. In technologically advanced and wealthy countries, the need for this element can be satisfied by the use of chemical fertilizers. Even in these regions, however, the cost of fertilizers is high. On the other hand, in the developing counties, fertilizers and specifically nitrogen-containing fertilizers are either largely or wholly unavailable or are so expensive that they cannot be used by farmers, particularly the small farmers that account for much of the food produced in these regions. Because of the need for this element combined with the high costs or unavailability of fertilizers, nitrogen may be said to be the key factor, in addition to water in water-poor regions, that regulates the quantity and quality of food and that is responsible for undernutrition and malnutrition.

It has been known for more than a century that biological nitrogen fixation is one of the means of introducing that critical element into agricultural ecosystems. Free-living bacteria and blue-green algae as well as plant-microbial associations convert atmospheric nitrogen to forms that ultimately can be utilized by higher plants, either to species living in usual soil circumstances or to rice grown under water. Particular attention has been devoted to the association between members of the genus *Rhizobium* and legumes because legumes are important sources of protein for humans and animals. Where these microorganisms are particularly active, especially in association with legumes or in fields of rice planted under water, much of the nitrogen demand of the crop can be satisfied by microbial processes.

Unfortunately, the practical use of nitrogen-fixing microorganisms has often been ignored because of the rapid development of the fertilizer industry. Ignoring these microorganisms in the developing countries is especially unfortunate because of the ease of introducing the requisite technology and the unavailability and high cost of alternatives to the biologically-mediated nitrogen fixation. In recent years, fortunately, there has been a major effort to set

right the deficiencies in concern with nitrogen fixation. These
efforts are evident in the large amount of research and the development of new technologies that relate to the agronomic, genetic and
biochemical aspects of nitrogen fixation. However, apart from the
developed countries, few trained personnel are available to imple-
ment the findings of either the older investigations or of the most
recent studies. Most of the countries of Latin America, Asia and
Africa hava a paucity of agronomists, soil scientists, and microbiologists who are familiar with the concepts, techniques and utility
of those approaches that have proven to be so successful in Europe,
North America, Australia and New Zealand.

To help overcome the deficiency in trained personnel, a training
course was conducted in Caracas, Venezuela on the subject of "Biological nitrogen fixation and its ecological basis." The chapters of
this book represent the substantive material presented as part of
the lectures in this training course. The subjects were well received by the participants in that course, and it is hoped that, by
publishing the manuscripts derived from the lectures, more widespread use will be made of the information that was presented. In
this way, it is hoped that a contribution can be made in overcoming
the deficiency of trained personnel in areas of the world where few
people currently are doing basic or applied research or are initiat-
ing field programs to introduce biological nitrogen fixation into
agricultural systems.

Martin Alexander
Cornell University
November, 1983

CONTENTS

TAXONOMY AND METABOLISM OF Rhizobium

AND ITS GENETIC RELATIONSHIPS

Gerald H. Elkan

Department of Microbiology
North Carolina State University
Raleigh, N. C. 27650. U.S.A.

IMPORTANCE OF BIOLOGICAL NITROGEN
FIXATION AND LEGUMES

As we approach the end of the twentieth century, all evidence points to an increasing worldwide food and protein deficit. This deficit will result from rapid population increases, the high cost and scarcity of energy, scarcity of arable land, and climatic changes. As to the last item, climatic changes have caused a 10% increase in the amount of world desert areas over the past 10 or so years. Atmospheric changes in ozone and carbon dioxide levels, deforestation, environment pollution, and/or cyclic atmospheric changes have all been implicated, but the reasons for this trend are not yet clear.

World population is increasing at an annual rate of 2.5% in the developing nations and at a 0.9% rate in the developed nations. Thus, in 1930 the world contained 2 billion (10^9) people; by 1960 this had grown to 3 billion, while today the world contains 4.4 billion people, three-fourths of whom live in the developing countries. If the current population trends continue, our earth will be inhabited by around 11 billion people by the year 2020, just 40 years from now. It has been determined that perhaps as many as 16 billion people could be supported eventually. The problem, however, is that this increase is coming about so rapidly and unevenly.

In the developing countries, total food production is increasing at a phenomenal 2.7% per year. However, the rapidly increasing population more than wipes out this gain so that there is an annual net loss in per capita food availability.

1

Today there is not really a world food shortage but rather a distribution problem. A few countries, mainly the United States, Canada, Australia, and New Zealand, are producing enough excess food to offset the shortage in the third world countries. Even this is a temporary situation, since while total world food production (in the developed and developing countries) is increasing at an approximate 3% annual rate, within 20 years we would have a doubling of food production annually to feed the world's people at that time.

How then do we increase production at a more rapid pace? Much of the better agricultural land is already in use (about 50% of the arable land). So land is scarce! In areas where the major unused land resources are located, i.e., Africa, South America, and Australia, we must first overcome a number of constraints such as the need for high energy-requiring technology, the need for irrigation, human disease problems, lack of population in the places where the land is located, lack of transportation, etc. While we can eliminate some of these constraints, the major increase needs to result from higher productivity per acre.

When we speak of the "food shortage," in the main we are concerned with energy and protein, or more precisely, essential amino acids. There are, of course, many other components of food, but energy and protein are the main world-wide problems. In this paper, we will concentrate on the protein problem. Where, then, are the main sources of available food protein? Seventy percent of the world's protein needs, and virtually all of the developing countries' needs, come from plant sources, mainly grain and secondly legumes. Only 30% of our protein comes from meat, milk, or eggs, although these are the major sources of protein in the developed nations.

If agricultural production is to be increased, we must then increase the yields of these sources of protein. Animal production requires feed. This feed is composed of grass, grain, or legumes alone or in combination. The scarcity of arable land is forcing more permanent grassland to become cultivated. Therefore, grasses will not be more plentiful. As noted earlier, grain and legumes are a direct major source of protein for man. Why use an animal, such as a steer, as an intermediary? As a matter of fact, this is quite wasteful. For example, if one takes the annual yield of soybeans from one acre and feeds beef cattle with the total harvest, the increased body protein from the animals would supply the needs for one adult for 77 days. Utilizing the soybean protein directly as food for humans would mean that the same adult would receive enough protein to last for six years.

With our present technology, we have been able to increase fish catches yearly for the past 20 or so years. Several years ago, these fish yields peaked and then decreased, apparently because of the overfishing of the known fishing grounds. The yields have since re-covered, but the lesson seems to be that there is a finite limit to the amount of fish harvest.

It seems unlikely, therefore, that animal protein supplies can be greatly increased by current technology, and given the scarcity of additional arable land, per acre yields on existing farm land of grains and/or legumes must be increased. Hopefully in the future, novel protein sources and synthetic proteins will be available, but at present we must depend upon conventional agriculture and plant protein. This premise resulted in the "green revolution."

The traditional grain cultivars (mainly rice, wheat, and barley) in the developing countries were tall-growing, low-yielding, and dis-ease and insect resistant. These crops also did not respond to ni-trogen fertilizer. That is, in the presence of added nitrogen, the stems grew longer until the plant lodged, but the seed head did not increase. Thus, no yield increases occurred. Crop breeders de-veloped a series of dwarf cultivars whose yield was increased by ni-trogen fertilizers and, in fact, required chemical nitrogen to give any yield. The plants were generally more insect and disease sus-ceptible, so insecticides and other pesticides were required. Thus, the green-revolution crops were dependent upon the plentiful use of cheap chemicals.

Yield increases were dramatic. Farmers are traditionally thought of as conservative. But in reality, when a useful develop-ment occurs, farmers will adapt very rapidly. The green-revolution acreage in 1965 totaled 200 acres, and just five years later, in 1971, 50,500,000 acres were harvested. For the first time, tradi-tional importing countries, such as India, were able to feed their burgeoning populations with home-grown food.

Most insecticides and herbicides have a hydrocarbon base. Ni-trogen fertilizer is produced from inert nitrogen gas from air. In the presence of heat and a catalyst (the Haber-Bosch process), hy-drogen from crude oil or natural gas is combined with the nitrogen to produce ammonia. From one ton of hydrocarbon, approximately two tons of ammonia are produced. It follows, therefore, that the cost of nitrogen fertilizer (and also of insecticides and herbicides) is dependent on the cost of hydrocarbons.

The increased use of agricultural chemicals caused competition for sources, driving costs up somewhat. In 1973, however, the en-ergy crisis drastically and suddenly increased the costs of these chemicals. For example, in 1971 one ton of urea delivered to the farm in the Philippines cost U.S. $45. In 1973 one ton of the same product cost U.S. $350. The cost has since gone down to about U.S. $125 but remains three-fold higher than before the crisis.

When the costs of these chemicals increased so suddenly, the cost of food to the consumer increased in the developed countries. However, in these countries 16 to 23% of the consumers' disposable income is used to purchase food. In the developing countries, this figure is 40 to 75%. The farmers, therefore, were not able to pass their increased costs onto the consumer. As a result, they purchased less nitrogen fertilizer with a concomitant yield decrease. In order to get optimum yields from the green revolution crops, in 1980 the farmers in the developing countries required 27 million metric tons of nitrogen fertilizer, but only about 7.5 million metric tons were utilized.

During the 19th century, in the United States, a farmer required about one calorie of energy input to produce one food calorie. With the intensive, highly efficient agricultural production in this country, it requires almost ten calories of energy input to produce one calorie of food. Labor and fuel energies are relatively unimportant. Rather, the energy input from agricultural chemicals, irrigation, equipment production, etc., all contribute heavily. Relatively small yield increases require even larger energy inputs. For example, in 1940, the average nitrogen fertilizer additions per acre of wheat grown in the United States were six pounds. In 1974 an average of 125 pounds were added per acre of wheat. This 21-fold increase in nitrogen fertilizer usage doubled the average yield of wheat per acre.

If the developing countries cannot afford the cost of the energy required, then they must develop a lower energy-requiring agricultural regimen. The highest energy inputs are required for animal production (cattle, poultry, fishmeal) with the exception of range-fed beef. Most plant protein requires considerable energy input but less than for animal protein. Because of the high amount of protein produced by these plants, legumes are among the lowest energy-input requiring intensive crops. We saw earlier the high quantity of protein produced by a legume such as soybeans. The protein is also relatively high in essential amino acids except for the sulfur amino acids, cystine and methionine. These are readily available from other plant sources, however. Of additional prime importance is the fact that associated with these legumes (soybeans, peanuts, cowpeas, beans, peas, etc.) are nitrogen-fixing bacteria called rhizobia. Atmospheric nitrogen is inert as far as animals and higher plants are concerned. Only some few bacteria (including blue-green algae) have the ability to fulfill their nitrogen requirements by converting atmospheric nitrogen to a usable form and then synthesizing their amino acids and proteins. In the case of rhizobia, which live in nodules on the roots of legumes, the energy for this process comes from the host-plant photosynthate. In return, the excess fixed nitrogen is made available to the host plant. As much as 200 pounds of N per acre is fixed by these bacteria. We

are thus able to grow a crop of soybeans or peanuts without chemical nitrogen fertilizer. Growing legumes then requires a very low
energy input because (a) the nitrogen fertilizer is supplied by bacteria (and not hydrocarbons) and (b) the unit protein production is
so high. We have developed, in addition to the conventional con-
sumption of these foods, processes able to supply additional supple-
ments of protein, such as texturized vegetable protein. By exploiting the legume-Rhizobium symbiosis to its optimum, we have practical, low energy-requiring, quality protein sources which are usable
during the interim period until the time when synthetic proteins become available.

TAXONOMY OF Rhizobium

     Beneke (1912) described a bacterial species as merely a col-
lection of more or less similar clones that the taxonomist ties into
a bundle which he chooses to call a species. How small or large
this bundle is has depended upon the scientific insight of the individual taxonomist. The same principle has been applied to larger
bundles that are called genera and families. The present state of
bacterial taxonomy still largely depends upon this approach, even
though we are now beginning to learn something of the phylogeny of
bacteria (Fox et al., 1980; DeLey, 1974).

     Bergey's Manual of Determinative Bacteriology (Buchanan and
Gibbons, 1974) describes the family Rhizobiaceae Conn 1938 as consisting of normally rod-shaped cells without endospores. The cells
are aerobic, gram negative, motile, and have either one polar or
subpolar flagellum or two to six peritrichous flagella. Many carbohydrates are utilized, with a usual production of considerable extracellular slime. There are two genera: (a) Agrobacterium, all
species of which, with the exception of A. radiobacter, incite
cortical hypertrophies on plants, and (b) Rhizobium, the species of
which form nodules on the roots of Leguminosae.

     As described in Bergey's Manual of Determinative Bacteriology,
the genus Rhizobium consists of two groups of species plus a miscellaneous grouping. The species and their major characteristics
are summarized in Table 1. The current classification recognizes
six species based upon host infectivity coupled with certain biochemical tests. This classification is historically based on the
work of Baldwin and Fred (1927, 1929), Eckardt et al. (1931), and
Fred et al. (1932). These workers proposed 16 cross-inoculation
groups. Six of them were evaluated to species level as currently
listed in Bergey's Manual of Determinative Bacteriology. The assumption that each species of Rhizobium nodulates only plants within
a specified cross-inoculation group, and that within such a group
all rhizobia from a host can inoculate all other numbers of the

TABLE 1. Species of Rhizobium and Their Characteristics after Bergey's Manual

Species	Serum zone in litmus milk	Acid reaction in litmus milk	Growth rate	Host nodulated	Common name of group
Group I: Two to six peritrichous flagella; GC content of the DNA 59.1-63.1 mole % (T_m)					
Rhizobium trifolii	+	-	Fast[a]	Trifolium	Clover
Rhizobium leguminosarum	+	-	Fast	Pisum, Lens, Lathyrus, Vicia	Pea
Rhizobium phaseoli	+	-	Fast	Phaseolus vulgaris	French bean
Rhizobium meliloti	+	-	Fast	Melilotus, **Medicago**, Trigonella	
Group II: Polar or subpolar flagellum; GC content of the DNA 61.6-65.5 mole % (T_m)					
Rhizobium japonicum	-	-	Slow[b]	Glycine max	Soybean
Rhizobium lupini	-	-	Slow	Lupinus, Ornithopus	Lupine
Rhizobium spp.	Variable	Variable	Variable	Vigna, Desmodium, Arachis, Centrosema, Stylosanthes, etc.	Cowpea miscellany

[a] Colonies circular, convex, semitranslucent, raised, and mucilaginous; usually 2 to 4 mm in diameter on yeast extract-mannitol-mineral salts agar (YEM) within 3 to 5 days (turbidity in YEM broth in 2 to 3 days).

[b] Colonies circular, punctiform, opaque, rarely translucent, white, convex, and granular; do not exceed 1 mm in diameter within 5 to 7 days on YEM agar (turbidity in YEM broth in 3 to 5 days or longer).

group, has lost credibility. Infact, D. C. Jordan and O. N. Allen
(Buchanan and Gibbons, 1974) state that the taxonomic position of
Rhizobium is controversial and that the current classification can
be regarded as only tentative. They conclude that a classification
of Rhizobium based on the cross-inoculation (or plant affinity) concept, because of widespread anomalous cross-infections, is not very
satisfactory. Graham (1976) summarizes the major limitations of the
cross-inoculation concept of classification as follows:

(a) Cross-infection. Cross-infection is the nodulation of
    plants from one affinity group by rhizobia from another.
    Each of the species has now been shown to be cross-infec-
    tive to some degree.

(b) Insufficient nodulation data. It is stated in Bergey's
    Manual of Determinative Bacteriology that, of the 14,000
    or so known species of legumes, only 8 to 9% have been ex-
    amined for nodules and only 0.3 to 0.4% have been studied
    with respect to their symbiotic relationships with nodule
    bacteria.

(c) Scarcity of biochemical data. Many of the biochemical
    studies involve only a few strains, and so it is diffi-
    cult to generalize as to the taxonomic meanings of bio-
    chemical differences. One taxonomic difference that does
    appear real is the designation of two groups of Rhizobium
    based upon growth rate.

The genus Rhizobium has traditionally been divided into two
groups, as suggested by Löhnis and Hansen (1921). According to
Allen and Allen (1950), the term "fast growers" commonly designates
the rhizobia associated with alfalfa, clover, bean, and pea because
in culture these grow much gaster (less than one-half the doubling
time) than the "slow growers," exemplified by soybean, cowpea, and
lupine rhizobia. As a result of an extensive study of acid production involving 717 strains of Rhizobium, Norris (1965) hypothesized
that the ancient form of the symbiont is represented by the slow
growers which produce alkali and are commonly associated with legumes
of tropical origin. Disputing this hypothesis, Graham (1964a) con-
tended that the differences between slow- and fast-growing Rhizobium
were too great to be based solely on evolutional differentiation of
root-nodule bacteria from an organism similar to the present-day
slow-growing type. Differences in carbohydrate nutrition between
fast and slow growers have been shown by many workers. In their review, Allen and Allen (1950) stated that the slow-growing rhizobia
were more specific in their carbohydrate requirements in every re-

spect. The great difference in carbohydrate utilization between
fast- and slow-growing root-nodule bacteria was confirmed by Graham
(1964b), who concluded that tests of carbohydrate utilization were
clearly valid criteria for the subdivision of Rhizobium. For example, Rhizobium trifolii and R. leguminosarum can utilize 20 different carbohydrates and tricarboxylic acid cycle intermediates for
growth, whereas R. japonicum can utilize only 8 of the 20.

The relative fastidiousness of the slow growers has been substantiated by more recent studies (Elkan and Kwik, 1968; Graham,
1964a, b). Specific enzymes may differ in the two groups; for example, Martinez-de-Drets and Arias (1972) proposed an enzymatic
basis for differentiation between these groups based on the presence
of an NADP-6-phosphogluconate dehydrogenase found in fast growers
but absent in slow growers. While the major biochemical pathways
seem to be similar, there is evidence that the preferred pathways
may be different. However, most of the metabolic studies were done
with only one or very few Rhizobium strains (Elkan and Kuykendall,
1981).

Based on the great similarities between the species of fast-
growing rhizobia and equally great differences between them and the
slow growers (such as R. japonicum), Graham (1964a) proposed the
consolidation of R. trifolii, R. leguminosarum, and R. phaseoli into
a single species and the creation of a new genus, Phytomyxa, to contain strains of slow-growing Rhizobium (type species: Phytomyxa
japonicum).

Vincent (1977) surveyed the literature and concluded that fast
or slow growth on yeast extract-mannitol agar, where the fast
growers lower the pH of the medium and the slow growers produce an
alkaline end point, is a major acceptable distinction between species of Rhizobium. This differentiation has also been utilized in
Bergey's Manual of Determinative Bacteriology (Table 1), but Vincent
includes in the fast growers some members of the cowpea miscellany.
The slow growers include two reasonably well-defined subgroups,
R. lupini and R. japonicum, plus a large number of relatively poorly
defined members of the cowpea miscellany.

As the cross-inoculation approach became more unsatisfactory,
alternative taxonomic methods were applied to these organisms.
These involved nutrition, Adansonian analysis, serology, DNA base
ratios, DNA hybridization, and, more recently, protein fingerprint-
ing or analysis.

Graham (1964a) studied 100 features in an Adansonian analysis
and concluded

   (a) that R. leguminosarum, R. trifolii, and R. phaseoli should
        be united to form a single species,

(b)   that R. meliloti should be maintained as a separate species,
      and

(c)   that R. japonicum and R. lupini should be combined with
      organisms of the cowpea miscellany to form a single species
      and perhaps, with additional studies, a genus.

Moffett and Colwell (1968), in a different Adansonian analysis,
found substantial agreement with these findings of Graham. They
also substantiated the fact that the major separation of Rhizobium
species is into fast and slow growers and that this separation is
apparently at the generic level. Using a different method of numerical taxonomy, 't Mannetje (1967) reexamined the findings of
Graham (1964a). He proposed that Rhizobium be considered as consisting of two sections, one with R. japonicum as the type species
and the other with R. leguminosarum as the type species. He further
recommended that the genus Rhizobium not be split at the genus level
until much more information is available.

DeLey and Rassel (1965) examined the DNA base composition and
flagellation as taxonomic tools. Using 35 strains of Rhizobium,
they found a correlation between the DNA base composition and the
type of flagellation. The peritrichously flagellated fast-growing
rhizobia had a low percent GC composition, ranging from 58.6 to
63.1%. These organisms were found in all cross-inoculation groups
examined. The subpolarly flagellated, slow-growing strains ranged
from 62.8 to 65.5% GC. From this study and a review of the literature, they tentatively proposed that there were two subgroups of
fast growers (R. leguminosarum and R. meliloti), while the slow
growers consisted of one species (R. japonicum). Thus, they proposed three species within the genus Rhizobium.

DeLey (1968), in an extensive review, critically analyzed the
data from the various studies involving numerical taxonomy, DNA base
ratios, and DNA hybridization. Based on these data, he proposed the
revision of the genus Rhizobium shown in Table 2.

Graham (1969) reviewed the analytical serology of the Rhizobiaceae. Based on the studies of many workers, he proposed three
broad serological groups:

(a)   R. trifolii, R. leguminosarum, and R. phaseoli;

(b)   R. meliloti; and

(c)   R. japonicum and R. lupini.

These serological groupings are in agreement with the numerical
taxonomic groupings of Graham (1964a). DNA homologies of the above
groups of organisms (R. trifolii, R. leguminosarum, and R. phaseoli)

TABLE 2.   Species of Rhizobium and Their Characteristics[a]

Species	Relation to the current species	Flagellation	Percent GC	Serum zone in litmus milk	Acid reaction in litmus milk	Growth rate	Nodule-forming characteristics, special features
Rhizobium leguminosarum	R. phaseoli	Peritrichous	59.0-63.5	+	-	Fast	Forms nodules on one or more of Trifolium, Phaseolus vulgaris, Vicia, Pisum, Lathyrus, Lens
Rhizobium meliloti	Unchanged	Peritrichous	62.0-63.5	+	+	Fast	Forms nodules with Melilotus, Medicago, Trigonella
Rhizobium rhizogenes	A. rhizogenes	Peritrichous	61.0-63.0	+	-	Fast	Causes hairy-root disease of apples and other plants
Rhizobium radiobacter	A. tumefaciens + A. radiobacter + A. rubi	Peritrichous	59.5-63.0	+	-	Fast	Frequently produces galls on angiosperms; produces 3-keto-glycosides
Rhizobium japonicum	R. japonicum + R. lupini + cowpea miscellany	Subpolar	59.6-65.5	-	-	Slow	Nodulates many different legumes including one or more of Vigna, Glycine, Lupinus, Ornithopus, Centrosema, etc.

[a]After DeLey (1968).

were determined by Jarvis et al. (1980). Based on the data obtained, they proposed that R. trifolii and R. leguminosarum be combined into one species and that within this species various biovars be designated according to plant specificity. It was recommended that R. phaseoli be retained as a separate species pending more study.

Using three different methods of nucleic acid hybridization, Gibbons and Gregory (1972) concluded that R. leguminosarum and R. trifolii could not be distinguished. R. lupini and R. japonicum were also shown to be closely related, and a close relationship also appeared between R. meliloti and R. phaseoli (but with less certainty). In a recent study, Hollis et al. (1981) were not able to differentiate between the reference strains (in Bergey's Manual of Determinative Bacteriology) of R. japonicum and R. lupini by DNA–DNA hybridization.

Protein analysis by two-dimensional gel electrophoresis was used by Roberts et al. (1980) to classify Rhizobium strains. They reported that all the slow-growers (group II) were closely related and were quite distinct from the group I isolates. R. meliloti strains, as previously shown using other methods, formed a distinct group, while the rest of the group I organisms seemed to be more diverse. All the previously discussed studies were devoted to taxonomic relationshps among species of Rhizobium. In many cases, because of the number of species (plus the cowpea miscellany), there usually was a relatively small representation for each group or species.

Few studies have been concerned with the taxonomic status within a species. In one such study, Hollis et al. (1981) examined the genetic interrelationships, using DNA–DNA hybridization, of 29 diverse isolates of R. japonicum. The results indicated that R. japonicum strains fall into two widely divergent DNA-homology groups plus one subgroup. These two groups are well out of the limits defining a species and thus illustrate the diversity within this so-called species. Ligon et al. (1982) used DNA–DNA hybridization to study the genetic taxonomy of the "cowpea miscellany" rhizobia. The cowpea rhizobia are a diverse group of promiscuous organisms which traditionally have not been included in the cross-inoculation system. The host range of the cowpea rhizobia is broad and includes a variety of tropical legumes, including the agriculturally important peanut (Arachis hypogaea) and cowpea (Vigna unguiculata). The results indicated that the cowpea rhizobia are indeed genetically very diverse and that most of the isolates studied did not belong to the same species; and in fact, the data suggested that probably several genera were represented.

Graham (1976) expressed his great concern about the decline in numbers of taxonomic papers on Rhizobium published since 1968. At

TABLE 3.   Skeletal Form of the Proposed New Taxonomic Treatment of
           the Rhizobiaceae

Kingdom Procaryotae	High Taxonomic Categories
Division II	The Bacteria
Part 7	Gram-negative aerobic rods and cocci
Familly III	Rhizobiaceae
Genus I	Rhizobium
Genus II	Bradyrhizobium
Genus III	Agrobacterium
Genus IV	Phyllobacterium

GENUS I: Rhizobium

R. leguminosarum
  biovar trifolii
  biovar phaseoli
  biovar viceae

R. meliloti

R. loti - fast-growing, sub-polar flagellated strains from Lotus and
          Lupinus with strong affinity for L. corniculatus, L. densi-
          florus, and Anthyllis vulneraria (but also nodulates Orni-
          thopus sativum). Includes the fast-growing strains nodu-
          lating Cicer, Sesbania, Leucaena, Mimosa, and Lablab.

GENUS II: Bradyrhizobium

Slow-growing, polar or sub-polar flagellated strains nodulating soybean, Lotus uliginosus, L. pendutulatus, and Vigna. Includes those
slow-growing strains nodulating Cicer, Sesbania, Leucaena, Mimosa,
Lablab, and Acacia. The possibility exists that other species will
eventually be defined within this genus, but for the present it is
suggested that, other than B. japonicum (the type species), the
various cultures be designated as Bradyrhizobium sp. with the trap
plant being designated ex. Bradyrhizobium sp. (Vigna), Bradyrhizobium sp. (Cicer), et.

B. japonicum

present the situation has not been changed. In view of the ever-
increasing importance of Rhizobium, it appears vital that the taxonomy of this genus be clarified. Detailed studies within each of the
current species and especially the cowpea miscellany need to be carried out. There appears to be a dearth of biochemical and broad-
spectrum serological tests useful for classification and identification of rhizobia. More studies of nodulation, involving additional

plant species, need to be conducted. Only if considerably more of
such data are obtained can this taxonomy be properly codified.

An interim reorganization of the family Rhizobiaceae is being
proposed by the International Subcommittee on Agrobacterium and
Rhizobium of the International Committee on Systematic Bacteriology
of the International Union of Microbiological Societies (D. C.
Jordan, personal communication). This revision will be included in
the ninth edition of Bergey's Manual of Determinative Bacteriology.
It is proposed to split the rhizobia into two separate genera: (a)
Rhizobium (the fast-growing rhizobia) and (b) Bradyrhizobium (the
slow-growing rhizobia). Also, added to this family is the genus
Phyllobacterium (the leaf-nodule bacteria).

The genus Rhizobium will consist of three reorganized species:
R. leguminosarum, which will contain three biovars (biovar trifolii,
biovar phaseoli, and biovar viceae); R. meliloti; and R. loti. It
will be recognized that the reorganization combines into one the
former species of R. leguminosarum, R. trifolii, and R. phaseoli.
The fast-growing members of the cowpea rhizobia and the former
species R. lupinus have been included in species R. loti. The new
genus, Bradyrhizobium, is made up of one species, B. japonicum,
which consists of the former species R. japonicum plus the slow-
growing members of the cowpea rhizobia. Table 3 summarizes the proposed new classification for the Rhizobiaceae.

METABOLISM OF RHIZOBIUM

If the legume-Rhizobium symbiosis is to be exploited more effectively, more detailed information is needed as to carbon and nitrogen
metabolism and biochemical regulation of the bacterium. There are
two aspects of such studies which are essential if the system is to
be manipulated and optimized. Studies of both the biochemistry of
cultured organisms as well as the bacteroids in the symbiotic host-
cell association are needed.

Presently, there are considerable gaps in our understanding of
the carbon, energy, and nitrogen metabolism. In this review, we
will consider mainly the status of the metabolism of free-living or
cultured rhizobia.

Nitrogen Metabolism

The most important aspect of rhizobia, for us, is the presence
of the inducible enzyme nitrogenase, which enables these bacteria to
reduce atmospheric dinitrogen to ammonia.

The overall reaction of this process has been summarized by
Bergersen (1971) as follows:

$$N_2 + nATP + 6NADPH + 2H \rightarrow 2NH_4 + nADP + nP_i + 6NADP + 6e$$

where n = 6.0 to 6.9 or 6.5 ATP/NH$_4$ depending on whether cell-free or cell mass-balance figures, respectively, are used. The nature of the reductant varies in different nitrogen-fixing systems and has not been firmly established. The biological nitrogen-fixation system in Rhizobium is roughly twice as efficient as the Haber Bosch chemical process (Rawsthorne et al., 1980).

Ammonia produced by the activity of nitrogenase is usually synthesized into other organic compounds prior to transport to the plant host. It has been shown that nitrogen-fixing R. japonicum in culture can excrete ammonia directly (O'Gara and Shanmugan, 1976). There exist two general ammonium-assimilating enzyme systems, one catalyzed by glutamate dehydrogenase (GDH; EC 1.4.1.3) and the other by glutamine synthetase-glutamate synthase (GS; EC 6.3.1.2/GOGAT; EC 2.6.1.53). These pathways were described in enteric bacteria, and which pathway operates depends upon the available level of both ammonia and energy. Ludwig (1978) summarizes these reactions as follows:

With excess. ammonium, GDH reductively aminates 2-ketoglutarate to yield glutamate (see reaction 1 below), whereas, with limited ammonium, GS and glutamate synthase (GOGAT; L-glutamine:2-oxoglutarate aminotransferase, EC 2.6.1.53) operate in concert to yield the same net result (see reactions 2 and 3 below). However, in the latter case, 1 mol of ATP is hydrolyzed per mol of net glutamate formed.

$$\text{2-ketoglutarate} + NH_4 + \frac{GDH}{NADPH} \longrightarrow \text{L-glutamate} \tag{1}$$

$$\text{L-glutamate} + NH_4^+ + ATP \xrightarrow{GS} \text{L-glutamine} + ADP + P_i \tag{2}$$

$$\text{L-glutamine} + \text{2-ketoglutarate} \xrightarrow[NADPH]{GOGAT} \text{2L-glutamate} \tag{3}$$

An active glutamine synthetase appears to be required for symbiotic nitrogen fixation (Konderosi et al., 1977; Ludwig and Signer, 1977). Konderosi isolated a glutamine-requiring mutant from a fast-growing Rhizobium (R. meliloti). The mutant had only 5% of the GS activity of the parent strain and formed small white nodules that did not fix nitrogen. Ludwig and Signer (1977) used nitrosoguanidine treatment to produce glutamine-dependent mutants with low GS activity from the slow-growing cowpea Rhizobium 32H1. The mutants formed nodules but lacked nitrogen-fixing ability. Revertants regained their nitrogen-fixing activity.

In addition to fixing nitrogen, many strains of Rhizobium can utilize nitrate or ammonia as the sole or supplementary source of nitrogen. In these circumstances, care has to be taken to avoid an inhibitory change of pH in a lightly buffered medium, because of selective uptake of an anionic or cationic source of N (Vincent, 1977).

Strains unable to use inorganic combined N are generally satisfied by the addition of a single amino acid. Glutamic acid is generally acceptable, but some strains and substrains prefer other amino acids. A suitable carbon compound, such as α-ketoglutarate (Elkan and Kwik, 1968; Jordan & San Clemente, 1955; Bergersen, 1961) may replace the need for an amino acid, provided a source of combined inorganic N is also available. Vitamin-free casein hydrolyzate is superior to any tested combination of amino acids for the growth of R. japonicum, perhaps because of a peptide growth factor. Nitrogenase has recently been demonstrated in cultured rhizobia by the use of media containing carefully selected carbon sources and the establishment of a sufficiently reduced oxygen tension. Nitrate can be reduced either as an electron acceptor or as a source of assimilable ammonia. Some strains utilize urea.

Proteolysis, as judged by gelatin liquefaction, is weak, although the "serum zone" formed in milk by some species seems to involve some degree of proteolysis, as indicated by an increase in soluble N and amino N (Vincent, 1977).

R. meliloti has been shown to accumulate an internal pool of amino acid. This accumulation is energy dependent; movement from the pool depends on protein synthesis (Jordan, 1959). Transamination in rhizobia was also demonstrated with the synthesis by cell-free extracts of glutamic acid from α-ketoglutaric acid, when glycine, L-histidine, or D-alanine acted as $-NH_2$ donors (Jordan, 1953, 1955). There were no indications of specialized pathways with the cultured rhizobia.

High levels of asparate aminotransferase and alanine aminotransferase were detected in rhizobia grown in the presence of ammonium salts, as well as in bacteroids (Fottrell and Mooney, 1969). Concentration of these, as well as of glutamate dehydrogenase, was influenced by the N source in the growth medium. While each amino acid failed to stimulate the production of its own transferase, each enzyme was stimulated by ammonia, as well as, though to a lesser degree, by nitrate. The combination of NH_4^+ and α-ketoglutarate was best. Enzyme induction was inhibited by the inclusion in the medium of actinomycin D, puromycin, or cycloheximide.

Most of the studies with nitrogen metabolism have been done with bacteroids rather than with the cultured Rhizobium cells.

This is resulting in a significant knowledge gap when it comes to studies of genetics or regulation of nitrogen metabolism.

Energy Metabolism

The high metabolic energy cost of biological nitrogen fixation has been summarized by Evans et al. (1980), who calculated that a total of 28 moles of ATP are theoretically consumed in the reduction of one mole of N_2. Other workers have given different estimates, and the value is still in question. Although this high energy requirement is readily supplied by the catabolism of carbohydrates, photosynthesis and hence carbohydrate supply has been cited by a number of workers as a major limiting factor in nitrogen fixation by the Rhizobium-legume symbiosis. Thus, for the full exploitation of dinitrogen fixation, elucidation of carbohydrate metabolism in rhizobia is of importance.

Pioneering studies early in the century established the usefulness of glucose, mannitol, sucrose, and maltose as energy sources for the root-nodule bacteria, and it was concluded that no complex, unidentified substances were required for growth. These studies were followed by numerous investigations undertaken to determine the nutritional availability of various carbon compounds.

In a 1950 review by Allen and Allen (1), these authors stated that that "few groups of bacteria have been so thoroughly studied as have the rhizobia." Yet, much of the carbohydrate work has been directed toward nutritional studies designed for establishing growth conditions in culture or for taxonomic purposes. As to the pathways of carbohydrate metabolism, the information available to date is incomplete, fragmented, and sometimes contradictory. Recent investigations have employed mutant methodology. Duncan and Fraenkel (1979) have stated that "the general intermediary metabolism of Rhizobium has not been extensively studied." Ronson and Primrose (1979) also concluded that relatively little is known about the pathways of central carbohydrate metabolism in rhizobia. Recognizing the relative neglect of this most important topic, the purpose of this review is to bring together that which is known about carbohydrate metabolism in rhizobia.

Disaccharides. A number of early investigators (Zipfel, 1911; Fred, 1911-1912; Sarles and Reid, 1935; Neal and Walker, 1936; Georgi and Ettinger, 1941) showed the availability of disaccharides as energy sources for some rhizobia. Sucrose and maltose were the two most frequently reported, although sucrose also serves as an acceptable energy source. These papers reinforce the conclusion that the fast-growing strains of Rhizobium can utilize disaccharides as growth substrates, whereas the slow-growers certainly cannot grow well on these substrates.

Graham (1965) examined the carbohydrate response of 95 strains
of Rhizobium in seven species. Maltose, sucrose, and lactose were
included. While these disaccharides were utilized by 36 out of 40
fast-growing strains, only nine of 55 slow-growing strains utilized
these substrates. Elkan and Kwik (1968) examined the carbohydrate
response of 36 R. japonicum strains. None of these slow-growing
Rhizobium could utilize sucrose; 13 strains grew poorly on lactose,
none grew well, and 23 could not grow at all. Six grew well on
maltose, 23 did poorly, and seven could not utilize this substrate.
Shmyreva and Plaksina (1972) examined the ability of three strains
each of R. lupini and R. japonicum to grow on either sucrose, lactose, or maltose, and they found no growth of these slow-growing
Rhizobium strains with these substrates. They thus corroborated the
work of Elkan and Kwik (1968) and concluded that the inability to
utilize disaccharides was "one of the peculiarities of slow-growing
nodule bacteria." Martinez-de-Drets et al. (1974) further examined
the growth on sucrose by 64 fast and slow-growing Rhizobium strains
and also confirmed the difference in the abilities of these two
groups to utilize the disaccharides maltose, lactose, and sucrose.
For the metabolism of sucrose, an inducible invertase (sucrose hy-
drolase) was found in the fast-growing strains, but invertase was
not present in the slow-growing strains. The invertase found exclusively in the fast-growing species was induced by sucrose, lactose, or maltose but was at only a low constitutive level in cells
grown on either glucose or fructose. A glucosido-invertase was
present in all extracts and, in addition, some strains had a β-D-
fructofuranoside-invertase as well. No sucrose phosphorylase activity was found in either of the fast-growing or slow-growing strains.

The enzymatic basis for lactose metabolism in R. meliloti has
been reported by Neil et al. (1977). These workers reported two
different β-galactosidase (EC 3.3.1.23) activities present in strain
2011, one inducible and one low-level constitutive. The two enzymes
were distinguished by differing concentrations of ammonium sulfate
required for their precipitation from solution and were separable by
gel electrophoresis. The inducible enzyme was missing in a mutant
strain able to hydrolyze lactose only slowly. Ucker and Signer
(1978) studied the effect of succinate on lactose metabolism in
R. meliloti. They described a phenomenon similar to catabolite repression since succinate was found to repress β-galactosidase synthesis in R. meliloti. A lac mutant with unaltered symbiotic properties lacked β-galactosidase activity. Ucker and Signer (1978) also
described the isolation of a pleiotropic mutant, which was unable to
grow on a mixture of cellobiose and maltose, in which β-galactosidase
activity was not inducible. This mutant made relatively few nodules
and did not fix nitrogen, but revertants occurring at a high frequency regained both properties.

Polyols. Mannitol is the traditional carbon and energy source
used for the in vitro cultivation of Rhizobium (Elkan and Kuykendall,

1981). Yet, Elkan and Kwik (1968) found that only 21 out of 36 R. japonicum strains tested showed a good growth response with D-mannitol as the carbon source; 11 out of 36 strains did not utilize mannitol.

Wilson (1937) determined that polyols were oxidized by rhizobia in an unusual manner; that is, the metabolic rate was initially low and increased with time. This was an early indication that polyol metabolic enzymes are inducible. Burris et al. (1942) first interpreted and described the growth of Rhizobium on polyhydric alcohols as a result of adaptive or inducible enzymes. They found that the oxidation of polyols was much greater when the bacteria had been previously grown on the same substrate. Martinez-de-Drets and Arias (1970) investigated polyol metabolism in R. meliloti and found that this species possessed two distinct enzymes, one for the metabolism of D-mannitol and D-arabitol and one for D-sorbitol metabolism. The NAD-specific enzyme responsible for the dehydrogenation of D-mannitol and D-arabitol was a D-arabitol dehydrogenase, whereas a sorbitol dehydrogenase was found responsible for the oxidation of D-sorbitol. They also found that R. meliloti has an ATP-linked hexokinase that acts on fructose (fructokinase) and a phosphohexose isomerase (EC 5.3.1.9) that acts on fructose-6-phosphate, but they reported that a hexose isomerase capable of interconverting glucose and fructose was absent. Hornez et al. (1976) noted that fructose and glucose are metabolized via different pathways in R. meliloti. Fructose is directly phosphorylated and utilized through the Embden-Meyerhof-Parnas (EMP) pathway.

Kuykendall and Elkan (1976) isolated derivatives of USDA strain 110 of R. japonicum that differed in the ability to utilize D-mannitol as a carbon source as the basis for their isolation; these also differed about 20-fold in nitrogen-fixation activity with soybeans (Kuykendall and Elkan, 1976). The symbiotically competent strains could not utilize mannitol, whereas the other derivatives could. Using the derivatives cloned from USDA strain 110, these investigators examined the enzymatic basis for D-mannitol utilization in R. japonicum. The ability to utilize D-mannitol was determined by the presence of an inducible NAD-dependent D-mannitol dehydrogenase (EC 1.1.1.67) capable of using either D-mannitol or D-arabitol as substrate. D-mannitol also induced its own specific transport system in R. japonicum (Mulongoy and Elkan, 1978). In R. japonicum, the substrate of preference was D-mannitol over D-arabitol since specific activities were two-fold higher with D-mannitol. In contrast, Martinez-de-Drets and Arias (1970) found that the analogous enzyme in R. meliloti, a D-arabitol dehydrogenase, had three to five-fold higher activities with D-arabitol than with D-mannitol as substrate. This difference in enzymatic specificity between fast-growers and slow-growers may reflect significant evolutionary differences distinguishing these groups into distinct origins.

Mannitol has a profound effect on glucose metabolism in R. japonicum. Its presence in the growth medium stimulates the synthesis of the nicotinamide adenine dinucleotide (NAD)-linked 6-phosphogluconate dehydrogenase, and it represses the glucose-uptake system two to three-fold (Mulongoy and Elkan, 1978). Addition of D-mannitol to cell suspensions immediately results in an approximate 50% reduction in adenosine triphosphate (ATP) levels. This is thought to be at least partly a result of D-mannitol kinase activity, but it may represent a vestige of the system found in most bacteria in which mannitol initially is transported into the cell by an inducible phosphenol pyruvate (PEP)-dependent phosphotransferase system and is subsequently converted to fructose-6-phosphate by an inducible mannitol-1-phosphate dehydrogenase. An unknown component (resistant to hydrolysis) of R. japonicum extracellular polysaccharides is assumed to be a mannitol phosphate polymer (Keele et al., 1974). This conjecture warrants investigation to determine whether the mannitol kinase activity present in R. japonicum is responsible for the synthesis of precursors of complex and, as yet, unidentified polysaccharides.

At present, it is concluded that in slow-growing as well as fast-growing Rhizobium, mannitol metabolism begins with a dehydrogenation producing D-fructose, as it does in Pseudomonas aeruginosa, rather than by a phosphorylation of the free hexitol, as it does in Escherichia coli. The fructose is then phosphorylated via a nonspecific hexokinase to form fructose-6-phosphate, which then enters the Embden-Meyerhof-Parnas pathway (Kuykendall and Elkan, 1977).

Five different polyol dehydrogenases have been described in R. trifolii by Primrose and Ronson (1980); inositol dehydrogenase, specific for inositol; ribitol dehydrogenase, specific for ribitol; D-arabitol dehydrogenase, which oxidizes D-arabitol, D-mannitol, and D-sorbitol; xylitol dehydrogenase, which oxidizes xylitol and D-sorbitol; and dulcitol dehydrogenase, which oxidizes dulcitol, ribitol, xylitol, and sorbitol.

Ronson and Primrose (1979) corroborated the observation of Martinez-de-Drets and Arias (1970) that D-glucose represses D-mannitol utilization by R. meliloti. Glucose was observed to repress the polyol dehydrogenase activities induced by mannitol or inositol only 20 to 40% and also almost totally repressed activities induced by ribitol or ducitol. For this repression to occur, glucose has to be metabolized to at least glucose-1-phosphate since D-mannitol dehydrogenase synthesis was not repressed by glucose in a glucokinase mutant (glk) of R. meliloti (Bergersen, 1960).

Hexoses. The specific catabolic pathways for hexose utilization by Rhizobium are as yet only roughly defined. The first suggestion of a specific pathway for carbohydrate catabolism in Rhizobium was made in 1952 by Jordan, who presented evidence for the oxidation of glucose via the Embden-Meyerhof-Parnas pathway. Katznelson (1955) reported that pyruvate and triose phosphate could be produced from 6-phosphogluconate by cell-free extracts prepared from glucose-grown cells of R. phaseoli, R. meliloti, R. leguminosarum, and R. trifolii. On the course of hexose phosphate utilization, Katznelson (1955) concluded that pyruvate was formed via the Entner-Doudoroff pathway. In a subsequent report, Katznelson and Zagallo (1957) demonstrated the presence not only of enzymes of the Entner-Doudoroff pathway, but also of enzymes of the Embden-Meyerhof-Parnas and pentose phosphate pathways. Two key enzymes of the pentose phosphate pathway found in extracts of the fast-growing Rhizobium species were NADP-glucose-6-phosphate dehydrogenase (EC 1.1.1.49) and NADP-6-phosphogluconate dehydrogenase (EC 1.1.1.44). Tuzimura and Meguro (1960) reported the oxidation of α-ketoglutarate, fumarate, succinate, and fructose-1,6-diphosphate by whole cells of R. japonicum, indicating the presence of the Embden-Meyerhof-Parnas pathway and tricarboxylic acid (TCA) cycle. Glucose catabolism in R. japonicum was studied by Keele et al. (1969) using a radiorespirometric method and by assaying for key enzymes of the known major energy-yielding pathways. The data were consistent with the Entner-Doudoroff pathway, which appeared to account for 100% of the catabolism of glucose in growing cells of R. japonicum. Because no 6-phosphogluconate dehydrogenase (NADP) activity was detected, an active pentose phosphate pathway was apparently lacking. Hexokinase, glucose-6-phosphate dehydrogenase, transketolase, and an enzyme system (Entner-Doudoroff dehydratase and aldolase) that produced pyruvate from 6-phosphogluconate were found. The finding of transketolase activity suggested that essential pentose phosphates could be synthesized from fructose-6-phosphate by transketolase and transaldolase reactions. The $^{14}CO_2$ evolution pattern from specifically labelled pyruvate indicated the presence of an active TCA cycle.

Both the fast-growing and slow-growing Rhizobium species have been shown by many workers to possess the Entner-Doudoroff pathway. Jordan (1962) concluded that the Embden-Meyerhof-Parnas and pentose phosphate pathways were also present in Rhizobium, and Siddiqui and Banerjee (1975) found the key enzyme of the Embden-Meyerhof-Parnas pathway, fructose-1,6-diphosphate aldolase, in cell-free extracts of slow-growing and fast-growing species.

Keele et al. (1970) followed their study of glucose catabolism with a study of D-gluconate catabolism in R. japonicum. In search of additional support for their earlier findings showing the involvement of the Entner-Doudoroff pathway and TCA cycle as the sole catabolic pathways, they found a preferential degradiation indicating primarily the Entner-Doudoroff pathway. However, an NADP-6-phosphogluconate dehydrogenase activity was detected, thus eliminating the pentose phosphate pathway. Having found gluconate dehydrogenase activity present, they postulated that a nonphosphorylated ketogluconate pathway that enters the TCA cycle at α-ketoglutarate was an ancillary pathway for D-gluconate catabolism in R. japonicum.

Martinez-de-Drets and Arias (1972) extended the observation made by Keele et al. (1969) about the absence of NADP-6-phosphogluconate dehydrogenase in R. japonicum to a comparative study of the fast-growing and slow-growing Rhizobium species. They concluded that the presence of this enzyme constituted an enzymatic basis for distinguishing the two groups since only fast-growing species possessed it. Then, Mulongoy and Elkan (1972b) found that R. japonicum, although lacking an NADP-specific 6-phosphogluconate dehydrogenase, did have an NAD-linked enzyme acting on 6-phosphogluconate. Chromatography of the reaction mixtures with partially purified preparations of this enzyme indicated that a phosphorylated ketohexonic compound was produced. Thus, the enzyme possibly could initiate a new pathway distinct from either the pentose phosphate pathway or the hexose cycle. However, the data of Martinez-de-Drets et al. (1977) appear to show that the NAD-6-phosphogluconate dehydrogenase of R. japonicum is a decarboxylating enzyme. This enzyme system is clearly deserving of much further study, particularly in order to resolve these differences. On the one hand, this enzyme may initiate a new pathway for the metabolism of hexoses in R. japonicum. The other possibility is that of participation of the pentose phosphate pathway in hexose utilization.

Several investigators have examined the naturally occurring variation in symbiotic nitrogen fixation among wild-type isolates of Rhizobium for a relationship between carbohydrate utilization and efficiency in symbiotic nitrogen fixation. Georgi and Ettinger (1941) concluded that no differentiation of efficient and inefficient strains of Rhizobium was possible on the basis of utilization of various carbohydrates. Katznelson and Zagallo (1957) compared the abilities of effective and ineffective strains of R. meliloti, R. leguminosarum, and R. phaseoli to metabolize 6-phosphogluconate to pyruvate, to oxidize glucose-6-phosphate, and to carry out aldolase and phosphohexokinase reactions. They found no distinct relationship between these metabolic properties and symbiotic competence. Symbiotically competent strains did, however, oxidize succinate more rapidly than did noncompetent strains. The main limitation of the study was the few strains used, and the authors noted that "these results can only be considered significant if duplicated with a larger number of strains." Gupta and Sen (1965) conducted a study in which the N_2-fixation efficiency of a Rhizobium isolated from legumes of each of four species was compared to the extent of glucose utilization by those strains in vitro. They examined 40 isolates each from pea (Pisum sativum), fenugreek (Trigonella foenumgraecum), black gum (Phaseolus mungo), and lablab (Dolichos lablab) for differences in the glucose-consumption capacity and in phosphate utilization, and they found that N_2-fixing efficiency (N content of plants grown in a nitrogen-deficient medium and inoculated with a particular isolate) was positively correlated with both asymbiotic metabolic parameters but particularly with the extent of glucose utilization. Magu and Sen (1969) studied the respiration rate of 15 strains of R. trifolii and of 10 strains of R. leguminosarum when growing on glucose, maltose, and mannitol, and they compared these results with those of plant-inoculation experiments in which N_2-fixation efficiency was again estimated by the increase in nitrogen content compared to controls. They did not find any correlation between efficiency and respiratory rate on these carbon sources. They noted, however, that there was a stimulatory effect by glycine on the respiratory rate during glucose utilization and that this stimulation was greater in efficient strains than in inefficient strains. Petrova et al. (1974) reported higher levels of glucose-6-phosphate dehydrogenase in nodules formed by effective strains of R. lupini than those formed by ineffective strains. This was correlated with a lower concentration of free sugar in bacteroids.

The closely related derivatives of strain 3I1b110 of R. japonicum differing in nitrogen-fixing efficiency that were isolated by Kuykendall and Elkan (1976) also differed in glucose utilization. Whereas the efficient derivative grew faster on D-glucose than did the inefficient strain, the reverse was true on D-fructose, on which the inefficient strain grew more slowly. This shows that glucose and fructose are metabolized differently, and it suggests a relationship between efficient glucose utilization and efficient symbiotic nitrogen fixation. Both strains possess an active TCA cycle and both metabolize glucose simultaneously by the Embden-Meyerhof-Parnas and Entner-Doudoroff pathways. The efficient strain, however, uses the former pathway preferentially, whereas the latter pathway predominates in the inefficient derivative. It was hypothesized that the higher N_2-fixation efficiency of strain I-110 may be a result of its higher efficiency in glucose utilization. However, these two strains differ by the same order of magnitude in their ability to synthesize nitrogenase in vitro as in their association with soybeans when grown on carbon sources such as D-gluconate, which they metabolize equally (Upchurch and Elkan, 1977). The levels of in vitro nitrogenase expression are much less than 1% of that found in symbiotic cells in nodules, and conclusions based on in vitro expression may not be valid. Bergerson (1960), however, proposed that hexoses may be the main substrates used by bacteroids in vivo.

Ronson and Primrose (1979) used a mutant-methodology approach to carbohydrate utilization in R. trifolii and to the question of the carbon source supplied to the microsymbiont by the clover plant. Carbohydrate-negative mutants of R. trifolii were isolated and characterized. They found mutants defective in glucokinase (glk; EC 2.7.1.2), fructose transport (fup), and the Entner-Doudoroff pathway and pyruvate carboxylase (pyc; EC 6.4.1.1). All the mutants, including one double mutant (glk fup), formed an effective symbiosis on red clover, thus showing that neither glucose nor fructose is fed to the microsymbiont by the host plant. Incidentally, the inability to demonstrate phosphofructokinase (EC 2.7.1.11) activity in cell-free extracts of R. trifolii as well as the finding of only low fructose-1,6-diphosphate aldolase (EC 4.1.2.13) activity indicated that the Embden-Meyerhof-Parnas pathway was not physiologically important in the strain studied. The operation of the Entner-Doudoroff pathway was shown by the production of pyruvate from 6-phosphogluconate. Pyruvate carboxylase (EC 6.4.1.1) mutants were unable to grow on any carbon source except succinate; therefore, as in Agrobacterium sp. and Pseudomonas aeruginosa, R. trifolii requires pyruvate carboxylase for growth on hexoses, pentoses, and trioses.

Carbohydrate-negative mutants of R. meliloti have also been isolated and studied. Arias et al. (1979) isolated a phosphoglucose isomerase-negative (pgi) mutant of R. meliloti which gave little nitrogen fixation in association with the alfalfa host plant. The mutant strain was isolated following nitrosoguanidine mutagenesis and screening glycerol-grown cells for lack of ability to utilize mannitol. This mutant did not grow on mannitol, sorbitol, fructose, mannose, ribose, arabitol, or xylose but grew on glucose, maltose, gluconate, L-arabinose, and other carbohydrates. The mutant strain accumulated high levels of fructose-6-phosphate when fructose was present; this toxic accumulation prevented growth on available carbohydrates such as L-arabinose. A revertant which was selected for fructose utilization had regained phosphoglucose isomerase activity, wild-type phenotype growth on all carbohydrates, and symbiotic N_2-fixing ability. Duncan and Fraenkel (1979) selected a mutant of R. meliloti unable to grow on L-arabinose. This mutant also could not grow on acetate or pyruvate, and it lacked α-ketoglutarate dehydrogenase activity. In R. meliloti, fructose and glucose are metabolized differently (Hornez et al., 1976). α-Ketoglutarate accumulates in cells growing on glucose. Fructose is directly phosphorylated and utilized via the Embden-Meyerhof-Parnas glycolytic sequence. R. meliloti cells growing on fructose or mannitol produce much extracellular polysaccharide but not when growing on glucose or galactose.

The results of Arias et al. (1979) suggest that R. meliloti does not have a functional Embden-Meyerhof-Parnas glycolytic sequence and that glucose and gluconate metabolized primarily via the Entner-Doudoroff pathway but partly by the pentose-phosphate pathway. The inability of the pgi mutant to grow on ribose, xylose, and arabitol suggests that the metabolism of these pentoses occurs via the nonoxidative pentose phosphate pathway leading to fructose-6-phosphate. Growth occurs on L-arabinose, however, because this pentose is utilized by a nonphosphorylated pathway leading to α-ketoglutarate (Duncan and Fraenkel, 1979).

Pentoses. Many workers have reported that the preferred carbon
source for the slow-growing species represented by R. japonicum is
L-arabinose (Elkan and Kuykendall, 1981). Pedrosa and Zancan (1974)
studied L-arabinose catabolism in R. japonicum and found a pathway
similar to that in pseudomonads. In this pathway, L-arabinose is
first converted to 2-keto-3-deoxy-1-arabonate in three steps: (a)
dehydrogenation of L-arabinose to form L-arabonolactone, (b) hydrolysis of L-arabonolactone to L-arabonate, and (c) dehydration of L-
arabonate to form 2-keto-3-deoxy-L-arabonate. The 2-keto-3-deoxy-
L-arabonate is then cleaved by a specific aldolase to yield glycoaldehyde and pyruvate. Their finding of an active NAD/NADP-depen-
dent glycoaldehyde dehydrogenase suggested a significant role for
glyoxylate cycle enzymes.

L-Arabinose metabolism was also studied in R. meliloti by Duncan and Fraenkel (1979). They found the pathway for L-arabinose
metabolism in R. meliloti differed from that found by Pedrosa and
Zancan (1974) for R. japonicum in leading to α-ketoglutarate rather
than pyruvate and glycoaldehyde. They found that R. meliloti lacked
2-keto-3-deoxy-L-arabonate aldolase but possessed α-ketoglutarate
semialdehyde dehydrogenase.

TCA Cycle Intermediates. Organic acids that are intermediates
in the cycle are utilized by many microorganisms as carbon and energy sources. However, some microbes that possess a complete TCA
cycle fail to actively transport these compounds, and consequently
they cannot metabolize them. Thus, R. japonicum has an active TCA
cycle (Keele et al., 1969, 1970; Mulongoy and Elkan, 1977), but only
two strains of 36 examined (Elkan and Kwik, 1968) grew well in vitro
with succinate as a carbon source. None of the 36 utilized either
citrate or malate. Earlier, Graham (1964) studied the utilization
of carbohydrates by diverse isolates of different species of Rhizobium, and he observed that the fast-growing species gave excellent
growth on most carbon sources tested, including TCA intermediates,
whereas the slow-growing Rhizobium strains were much more restricted
in their choice of metabolizable substrates. Specifically, R.
trifolii and R. leguminosarum could utilize 20 different carbohydrates and TCA cycle intermediates, whereas R. japonicum could utilize only eight of 20. Proctor (1963) found relatively weak oxidation of TCA intermediates by Rhizobium compared to other carbohydrates. He studied, using manometric techniques, the oxidation of
more than 50 compounds by four distinct strains of Rhizobium from
Trifolium pratense, Lotus uliginosus, and Galega officinalis. Except for oxaloacetate, TCA intermediates were only weakly oxidized;
aconitate was extensively oxidized by one of the strains. Proctor
(1963) found very rapid oxidation of proline, leading him to suggest that this amino acid may be a "true substrate for the rhizobia
in nodules." Tuzimura and Meguro (1960) studied the oxidation of
various carbohydrates and TCA cycle intermediates by Rhizobium extracted from soybean nodules compared with R. japonicum grown on

either glucose or succinate. Cells derived from nodules actively oxidized TCA cycle intermediates but did not metabolize glucose or other hexoses (or sucrose or mannitol) except for fructose-1,6-diphosphate. Succinate-grown cells showed the same pattern. Only glucose-grown cells actively metabolized glucose. They concluded that "These facts suggest that the energy sources of Rhizobium in symbiotic state are organic acids but not carbohydrates." Katznelson and Zagallo (1957) examined the substrate-specific metabolic activity of effective and ineffective strains of fast-growing species; effective strains of Rhizobium oxidized succinate more rapidly than ineffective strain.

The enzymes of the glyoxylate cycle were examined in nodules and in free-living Rhizobium by Johnson et al. (1966). This study followed from the observations that soybean nodules and bacteria extracted from soybean nodules contain relatively large amounts of fatty acids, poly-β-hydroxybutyrate (PHB) is an important form of energy storage in Rhizobium, and utilization of PHB would be expected to result in formation of acetoacetate and acetyl-CoA. They surveyed Rhizobium isolates from bush bean, cowpea, lupine, soybean, alfalfa, red clover, and pea. Interestingly, malate synthetase activity was high in nodules of plant species in symbiosis with slow-growing rhizobia, but was barely detectable in nodules of plant species participating in symbiosis with the fast growers. Significant isocitrate lyase activity was lacking in bacteria derived from any type of nodule, indicating that the glyoxylate cycle does not operate in the symbiotic metabolism of Rhizobium. They therefore postulated a structural role for the fatty acids found in nodules. When oleate was utilized as the carbon source for in vitro cultivation of Rhizobium, significant levels of isocitrate lyase were found. Duncan and Fraenkel (1979) confirmed and extended these experiments. They assayed the glyoxylate cycle enzymes in a strain of R. meliloti and a α-ketoglutarate dehydrogenase mutant. Malate synthetase activity was present in cultures grown on all carbon sources, but isocitrate lyase activity was present only in cells grown on acetate; this showed that the glyoxylate cycle was probably used only in acetate growth but not for either pyruvate of L-arabinose metabolism.

Triose Metabolism. In the nutritional survey of 36 strains of R. japonicum reported by Elkan and Kwik (1968), several trioses were included. Twenty-nine of the strains utilized glycerol; only six strains used pyruvate as the sole energy and carbon source. Neither lactate nor propionate was utilized by any of the strains. Because whole cells were used in these studies, transport problems might be a factor in the inability to metabolize the latter compounds. Graham (1964) found that pyruvate, the only triose included in his study, could serve as the sole carbon source for 53 of 95 strains representing seven Rhizobium species. Representatives of the fast-growing species of Rhizobium as well as R. japonicum were shown by Johnson

et al. (1966) to oxidize acetate and pyruvate and yet be unable to
utilize these compounds for growth. They suggested that this was
the result of an inability of the organism to produce four-carbon
intermediates from acetate because of the absence of isocitrate
lyase (EC 4.1.3.1).

Arias and Martinez-de-Drets (1976) examined glycerol metabolism
in four strains of R. japonicum and one strain of R. trifolii, all
of which could grow on glycerol. Cell-free extracts of glycerol-
grown cells of both slow-growing and fast-growing rhizobia contained
glycerol kinase (EC 2.7.1.30). This enzyme, which is specifically
induced by glycerol, catalyzes the phosphorylation of glycerol to
glycerophosphate, is located in the soluble fraction of the cell
extract, and requires ATP. No phosphoenol pyruvate phosphotransferase activity was found in the different fractions of the extracts. A glycerophosphate dehydrogenase catalyzing the oxidation
of glycerophosphate to dihydroxyacetone phosphate was detected in
the five Rhizobium extracts. This enzyme was found in the particu-
late fraction. No glycerol dehydrogenase activity was detected.
These data show that in Rhizobium, glycerol is metabolized through
the phosphorylated pathway similar to E. coli, Aerobacter aerogenes,
and Rhodopseudomonas capsulata (Elkan and Kuykendall, 1981). However, unlike E. coli, Rhizobium cannot grow on glycerophosphate.
The pathway for glycerol utilization does not appear to differ in
fast- and slow-growers, but the authors were careful to state that
more strains and species need to be examined before a definitive
conclusion is reached.

Since rhizobia have pyruvate carboxylase as well as an active
TCA cycle, pyruvate is utilized via the TCA cycle. An alternate
role for pyruvate has been described by Trinchant and Rigaud (1974).
Cell-free extracts of R. meliloti contained soluble lactate dehydrogenase (EC 1.1.1.27). This enzyme is found to catalyze the reduction of pyruvate to lactate in the presence of NADH. In addition, the enzyme reduces indole-3-pyruvic acid to indole-3-lactic
acid. Earlier these authors had found an alcohol dehydrogenase
(EC 1.1.1.1) that carries out the reduction, using NADH, of indole-
3-acetaldehyde to form tryptophan. A role of these enzymes and
pyruvate in indole metabolism is proposed.

De Hertogh et al. (1964) showed that propionate is oxidized by
bacteroids from soybean nodules and by cells of R. japonicum and
R. meliloti. They demonstrated the presence of enzymes required for
the metabolism of propionate by conversion to succinate via methyl-
malonate and then oxidation of succinate via the TCA cycle. Cell-
free extracts of R. meliloti and R. japonicum and soybean bacteroids
showed a capacity to catalyze the activation of propionate to propionyl-CoA when supplied with propionate, ATP, and coenzyme A. The
carboxylation of propionyl-CoA to form methylmalonyl-CoA via pro-

pionyl CoA carboxylase was also demonstrated. Radiorespirometric
experiments with specifically labeled propionate showed patterns consistent with the utilization of propionate by a series of reactions
resulting in the formation of succinate, which is then oxidized via
the TCA cycle.

<u>Summary</u>

The mechanisms of carbohydrate metabolism in <u>Rhizobium</u> are particularly deserving of research effort, since the efficiency of energy derivation from the utilization of carbon sources is a determining factor in the efficiency of symbiotic nitrogen fixation (Hardy
and Havelka, 1976). The literature available on carbohydrate metabolism in root-nodule bacteria provides information that is incomplete, fragmented, and sometimes contradictory. Some generalizations, however, are possible. Key enzymes of the Embden-Meyerhof-
Parnas, pentose phosphate, and Entner-Doudoroff pathways have all
been demonstrated in the fast-growing rhizobia, although a large
survey has indicated that fructose-1-6-diphosphate aldolase (critical for the Embden-Meyerhof-Parnas pathway) was relatively insigni-
ficant and sporadic. The presence or absence of 6-phosphogluconate
dehydrogenase is an important point of departure between the fast
and slow growers because it constitutes the essential link between
the Entner-Doudoroff and phosphate pentose pathways. With the vir-
tual absence of this enzyme, the slow-growing rhizobia are left essentially dependent on the Entner-Doudoroff glycolytic sequence.

R. japonicum has been investigated in some detail as a representative slow grower. The results substantiate its dependence on
the Entner-Doudoroff pathway when using glucose and largely, though
not entirely, on this system when supplied with gluconate. In the
second case, there appears to be a supplementary "<u>Acetobacter</u>-like"
direct oxidation of gluconate via 2-keto- and 2,5-diketogluconate,
then as α-ketoglutarate into the TCA cycle. This supplementary pathway has been suggested as a means of securing four-carbon compounds
when an absence of isocitrate lyase apparently prevents the usual
production from acetate.

Polyol entry into the glycolytic cycle has been demonstrated
as a result of inducible dehydrogenases by which mannitol and sorbitol produce fructose and arabitol gives rise to xylulose.

Other sugars probably are converted to fructose-6-phosphate
(mannose via mannose-6-phosphate; pentoses by the C_2-C_4 split) or to
glucose-6-phosphate (directly from glucose by means of the UDP-con-
trolled epimerization of galactose). Disaccharides require hydrolysis or phosphorolysis. An alternative pathway to pyruvate and gly-
ceraldehyde-3-phosphate (analogous to the Entner-Doudoroff sequence
for glucose) might also be available to <u>Rhizobium</u>. The block in
the pentose phosphate pathway characteristic of the slow growers

TABLE 4. Summary of Known Aspects of Carbohydrate Metabolism in _Rhizobium meliloti_ (fast-growing) and _Rhizobium japonicum_ (slow-growing)

Aspect of carbohydrate metabolism	_R. meliloti_	_R. japonicum_
Nutritional availability of various carbon compounds	Very versatile; e.g., disaccharides such as sucrose are readily used as are TCA cycle intermediates	More restricted in choice of substrates for growth, e.g., non-utilization of disacharides and often of TCA cycle intermediates
Substrate preference of D-mannitol/D-arabitol dehydrogenase	D-arabitol	D-mannitol
Catabolite repression	Observed	Not yet observed
Presence of Embden-Meyerhof-Parnas and Entner-Doudoroff enzymes	Demonstrated	Demonstrated
Pentose phosphate pathway	NADP-6-phosphogluconate dehydrogenase activity present	Unclear, no NADP-6-phosphogluconate dehydrogenase, but the NAD-linked enzyme may/may not be decarboxylating
L-Arabinose catabolism	Pathway leads to α-keto-glutarate	Different pathway leading to glycoaldehyde and pyruvate phosphogluconate

could well explain the common superiority of pentose sugars for these rhizobia in that these sugars would be brought directly into the system. The difference in metabolism between the fast and slow growing rhizobia is summarized in Table 4.(see p. 29).

RHIZOBIUM GENETICS

The increased interest in biological nitrogen fixation has occurred at a time of a rapid development in the state of the art of microbial genetics. There has been a concomitant interest in Rhizobium genetics. There has been considerable progress made in several areas, including determining the role of indigenous plasmids in symbiotic functions, the genetics of Rhizobium bacteriophage, mapping, and chromosomal gene transfer.

Mutant Isolation

Recently, there has been increased interest in the use of mutants of Rhizobium for studies of nitrogen fixation and/or legume associations. Spontaneous mutants have been isolated from presumptively pure cultures of R. japonicum. Clones of strain 3I1b110 were isolated that differed from the parent culture by as much as 20-fold in nitrogen fixing ability (Kuykendall and Elkan, 1976). The absence of an inducible D-mannitol dehydrogenase (EC 1.1.1.67) was also noted (Kuykendall and Elkan, 1977). This mutant did not utilize glucose as efficiently as the parent strain because it preferentially used the Entner-Doudoroff pathway rather than the higher energy-yielding Embden-Meyerhof-Parnas pathway (Mulongoy and Elkan, 1977). Meyer and Pueppke (1980) isolated another mutant from strain 3I1b110 which failed to bind radioactively labeled soybean lectin. A number of metabolic mutants were similarly isolated from other Rhizobium strains (Upchurch and Elkan, 1978). These spontaneous mutants were compared to the parent clones using bacteriophage typing, serotyping, and DNA-DNA homology (Elkan and Kuykendall, 1981). The results showed that these mutants were indeed not different rhizobia but clonal substrains.

Such testing of spontaneous derivatives is necessary to rule out cultures carrying two different rhizobia. O'Gara and Shanmugan (1978), for example, reported slow-growing R. trifolii isolates that, unlike the parent, could nodulate soybean plants and not clover. It was later shown that their culture contained both fast- and slow-growing rhizobia, a mixture of R. trifolii and R. japonicum.

The occurrence and properties of clonal derivatives of Rhizobium do not appear to be uncommon and need to be investigated further because these appear useful for genetic and biochemical studies. An important lesson is to be learned from these observations; that is, the genetic purity of Rhizobium cultures cannot be assumed. Ideally,

isolates of clones from single cells should be used for definitive
genetic and biochemical studies. Unfortunately, this is usually not
practical. Chemically induced point mutants demonstrating the desired phenotype are usually easier to isolate than spontaneous mutants. However, there is a lower probability of spontaneous mutants
containing other mutations in genes that are not examined in the
screening process. These methods can aid in clarifying the relationship between biochemical and symbiotic properties. For example,
Maier and Brill (1976, 1978) made mutants of R. japonicum 61A76 with
nitrosoguanidine and were able to isolate five mutants, after screening over 2500 surviving clones, which would not fix nitrogen with
soybean plants. Several of the mutants were nonnodulating (Nod⁻)
and lacked a surface antigen typical of the parent strain. Also,
after making mutants with nitrosoguandine, Beringer et al. (1977)
isolated a number of nonfixing mutants (Fix⁻) of R. leguminosarum,
some of which were temperature sensitive; that is, they were Fix^+
on peas at one temperature and Fix⁻ at a higher plant-growth temperature. In these studies, two approaches were used to evaluate
mutants. Either direct screening for symbiotic defects of the survivors of mutagenesis is made on plants, or classes of mutants are
selected (using markers such as altered antibiotic-resistance patterns) and then these mutants are tested on plants.

Mutants defective in various cell-surface components have been
used to substantiate the role of the cell surface in the ability of
rhizobia to infect and nodulate specific hosts. Spontaneous mutants of R. leguminosarum have been isolated on the basis of having
reduced levels of extracellular polysaccharides (Napoli and Albersheim, 1980; Sanders et al., 1978). Such mutants usually do not
nodulate peas, and the severity of the effect is inversely proportional to the amount of extracellular polysaccharide made by the
mutant. This indicates that the surface component is important in
the nodulation process.

Biochemical explanations for the alterations in nodulation
effectiveness and nitrogen-fixing ability of such mutants is difficult at this time because our knowledge of the regulation and control of infection and nodulation is very limited.

A major use of mutants is for the purpose of gene mapping.
Spontaneous and chemically induced mutations leading to resistance
to various antibiotics have been isolated by many workers. Such
mutants are useful as marker strains for ecological studies. Chemical mutants obtained with mutagens such as nitrous acid, nitrosoguanidine, and ethyl-methanesulfonate have been used to begin the
construction of the genetic maps of R. meliloti and R. leguminosarum
(Meade and Signer, 1977; Konderosi et al., 1977a; Beringer et al.,
1978b; Konderosi and Johnston, 1981). A wide range of auxotrophic
mutants has been obtained. There is considerable species variabil-
ity in the mutagenicity of various chemicals. The reasons for this

are unknown. In the slow growing rhizobia, in fact, very few auxotrophic mutants have been reported.

Recently, techniques have been adapted for us with Rhizobium
that allow transposons to be used as mutagens. Transposons are
translocatable elements that are stretches of DNA which, as discrete
genetic and physical entities, can move from one position in a
genome to another position in the same or a different genome. These
particles include bacteriophages Mu-1 as well as translocatable antibiotic-resistance elements (Tn). An advantage of these elements is
that a transposon inserted into a gene not only mutates it but also
marks it with the phenotype (i.e., antibiotic resistance) determined by the transposon. Thus, by mapping the resistance marker,
the mutation site can be located. Using the transposon Tn 5, for
example, a number of Fix⁻ mutants of R. leguminosarum have been isolated following screening of strains in which the transposon had
been inserted into a transmissible plasmid known to carry genes
concerned with a symbiotic function (Buchanan-Woloston et al., 1980;
Johnston et al., 1978a; Brewin et al., 1980). The marked plasmid
could now be identified on elimination or transfer.

Gene-Transfer Systems

Transformation. In Rhizobium, three main types of gene transfer system have been reported; conjugation, transformation, and
transduction. Transformation has been extensively reported. However, in rhizobia no transformations have been developed to a level
required for fine-scale chromosomal analysis, and no transformation
of Rhizobium with plasmid DNA has been reported as yet. This is an
important limitation since plasmids are implicated in the symbiotic
process. Recently, transformation of Agrobacterium tumefaciens, a
member of the Rhizobiaceae, with plasmid DNA has been reported, so
it would appear that a similar system might operate in Rhizobium.

Conjugation. Studies of conjugal gene transfer in Rhizobium
have resulted in the detailed information on chromosomal mapping.
The recent studies have made use of antibiotic-resistance Pl plasmids originally isolated from Pseudomonas. These plasmids have a
wide host range in various Rhizobium species (Datta et al., 1971;
Beringer, 1974; Meade and Signer, 1977; Konderosi et al., 1977a;
Kuykendall, 1979; Beringer et al., 1980). Several of these Pl
class plasmids are useful in promoting chromosomal recombination in
rhizobia (especially RP4, R68, and R68.45) and have been used for
chromosomal gene mapping. Mapping in these studies involved crosses
between multiple antibiotic or auxotrophic strains with subsequent
selection for transfer of one allele. The selected recombinants
were checked for coinheritance of other alleles, and the coinheritance frequencies were inversely related to map distance. So
far, there does not appear to be anything greatly peculiar about
the Rhizobium maps; the genes for specific biosynthetic functions
are more dispersed than in the enteric bacteria.

Transduction. Studies of transduction have involved the fast-growing rhizobia, and most of these have been concerned with R. meliloti. In this species, Kovalski (1970) was able to get $\overline{12}$ of 21 temperate bacteriophages to transduce an antibiotic marker. Beringer et al. (1980) reviewed a number of studies of generalized transduction in both R. meliloti and R. leguminosarum. Transduction occurred at a high frequency using antibiotic markers. Transductants have also been reported to transduce parts of large plasmids and even whole plasmids in these two Rhizobium species. In transduction, two generalizations are evident. There have been no reports of transduction in slow growing rhizobia. In our laboratory, for example, we have isolated a number of temperate and virulent bacteriophages for R. japonicum. Even though we used a number of different transducing techniques, we were not able to demonstrate transduction. The second point of omission is that there has been no report of transduction of symbiotic properties. This may mean that not many symbiotic functions are located on chromosomes, but, of course, it is also harder to isolate such mutants, which can only be identified by using plants.

Plasmids. It has long been observed that the symbiotic properties of rhizobia are relatively unstable in laboratory cultures. A number of workers (Higashi, 1967; Van Rensburg et al., 1968; Dunican and Cannon, 1971; Parijkaya, 1973; Zurkowski et al., 1973) found that treatment of fast growing rhizobia with plasmid-curing agents such as acridine orange, acriflavin, or sodium dodecyl sulfate decreased the symbiotic properties of these bacteria, suggesting a link between Fix^+ and Nod^+ functions and plasmids. In a related study, Cole and Elkan (1973) found evidence suggesting plasmid-encoded antibiotic-resistance genes in R. japonicum. In all of these studies, the evidence for the presence of the plasmid was indirect. All species of Agrobacterium, also a member of the family Rhizobiaceae, apparently contain large plasmids with molecular weights in excess of 10^8. Techniques used to isolate large plasmids from agrobacteria were applied to Rhizobium with the result that a wide range of large plasmids has been isolated and sized from most of the strains of Rhizobium studied (Sutton, 1974; Klein et al., 1975; Tshitenge et al., 1975; Dunican et al., 1976; Zurkowski and Lorkiewicz, 1976; Olivares et al., 1976; Nuti et al., 1977). The procedures used to isolate plasmids have been reviewed by Denarie et al. (1981). In both Rhizobium and Agrobacterium, these large plasmids control infectivity and host range, but there is no evidence that a part of any Rhizobium plasmid can be transferred to the host-plant genome in the manner of the plasmid DNA of A. tumefaciens. It should be mentioned that Tichy and Lotz (1981) isolated extremely large plasmids (up to 600 Mdalton) from strains of R. leguminosarum newly isolated from the nodules of pea plants. They posulated that, in some cases, these large plasmids may be multimers of the smaller plasmids that were also found in these strains.

Many workers have also reported the widespread presence of variously sized small plasmids in Rhizobium, but most of the studies have concentrated on the large plasmids. The available information about the genetic functions located on Rhizobium plasmids is still scant, but it is possible to give some idea of the genetic functions associated with these plasmids.

(a) Nodulating ability. Casse et al. (1979) reported that when a strain of R. leguminosarum was cured of its smallest plasmid, the derivative could no longer nodulate peas. Similar observations were made by Zurkowski and Lorkiewicz (1978) and Lie and Winarno (1979), who were able to restore nodulating ability to Nod⁻ strains of R. leguminosarum by the introduction of the proper plasmid.

(b) Nitrogen fixation. It is not known what part of the nitrogen-fixing system of rhizobia is encoded for by plasmids, but clearly at least some of the structural nitrogenase genes are plasmid borne. This was recently demonstrated by Ruvkun and Ausubel (1980) and Nuti et al. (1979).

(c) Host specificity. A number of workers have been able to demonstrate a change of infectivity in Rhizobium by plasmids so that the Rhizobium derivative was able to nodulate legume cultivars in addition to their normal hosts (Higashi, 1967; Johnston et al., 1978a; Lie and Winarno, 1979).

(d) Bacteriocin production. A number of small plasmids have been identified with the production of bacteriocins by rhizobia (Brevin et al., 1980). This may have some implications to the competitiveness of some Rhizobium strains.

(e) Cell wall polysaccharide. Prakash et al. (1980) reported that a Nod⁻ derivative of R. leguminosarum that had been cured of its smallest plasmid apparently had an alteration of its extra-cellular polysaccharides as demonstrated by a rough colony morpholgy. In view of the fact that the hypothesis for the specificity of the Rhizobium-legume interaction is based upon the chemical character-istic of the Rhizobium cell surface, this observation of plasmid control becomes significant.

Summary

Techniques have been developed with other bacterial systems that are adaptable (or have been adapted) to Rhizobium genetics. Therefore, the appropriate technology exists for genetic studies of the nitrogen-fixing metabolism of rhizobia. However, not enough is known of the metabolism and taxonomy of these bacteria. Genetic studies of nitrogen metabolism have also been hampered severely by the difficulty of obtaining nitrogen fixation in the absence of nodules. Nitrogen fixation <u>ex planta</u> has been reliably demonstrated only in the slow growing rhizobia. On the other hand, most metabolic and genetic studies have been conducted with the fast growing rhizobia because this is a more convenient system for study. Studies of the molecular taxonomy of the rhizobia emphasize the genetic unrelatedness of the fast and slow growing groups. Thus, it is difficult to extrapolate from the existing studies.

REFERENCES

E. K. Allen and O. N. Allen, Bacteriol. Rev., 14:273-330 (1950).
A. Arias, C. Cervenansky, A. Gardiol, and G. Martinez-Drets, J. Bacteriol., 137:409-414 (1979).
A. Arias and G. Martinez-Drets, Can. J. Microbiol., 22:150-153 (1976).
I. L. Baldwin and E. B. Fred, J. Bacteriol., 17:141-150 (1929).
I. L. Baldwin and E. B. Fred, Soil Sci., 24:217-230 (1927).
W. Beneke, Bau und Leben der Bakterien, Berlin Univ., Leipzig (1912).
J. E. Beringer, J. Gen. Microbiol., 84:188-198 (1974).
J. E. Beringer, N. J. Brewin, and A. W. B. Johnston, Heredity, 45: 161-186 (1980).
J. E. Beringer, A. W. B. Johnston, and B. Wells, J. Gen. Microbiol., 98:339-343 (1977).
J. E. Beringer, S. A. Hoggan, and A. W. B. Johnston, J. Gen. Microbiol., 104:201-207 (1978).
N. J. Brewin, J. E. Beringer, A. V. Buchanan-Wollaston, A. W. B. Johnston, and P. R. Hirsch, J. Gen. Microbiol., 116:261-270 (1980).
N. J. Brewin, J. E. Beringer, and A. W. B. Johnston, J. Gen. Microbiol., 120:413-420 (1980).
R. E. Buchanan and N. E. Gibbons, Bergey's Manual of Determinative Bacteriology, 8th ed., Williams and Wilkins, Baltimore (1974).
A. V. Buchanan-Wollaston, J. E. Beringer, N. J. Brewin, P. R. Hirsch, and A. W. B. Johnston, Mol. Gen. Genet., 178:185-190 (1980).
R. H. Burris, A. S. Phelps, and J. B. Wilson, Proc. Soil Soi. Soc. Am., 7:272-275 (1942).
F. Casse, C. Boucher, J. S. Julliot, M. Michel, and J. Denarie, J. Gen. Microbiol., 113:229-242 (1979).
M. A. Cole and G. H. Elkan, Appl. Environ. Microbiol., 37:867-870 (1979).
N. R. Datta, W. Hedges, E. J. Shaw, R. B. Sykes, and M. H. Richmond, J. Bacteriol., 108:1244-1249 (1971).
J. DeLey, Annu. Rev. Phytopathol., 6:63-90 (1968).
J. DeLey, Taxonomy, 23:291-311 (1974).
J. DeLey and A. Rassel, J. Gen. Microbiol., 41:85-91 (1965).
J. Denarie, P. Boistad, and K. Case-Delbart, in: Biology of the Rhizobiaceae, (K. L. Giles and A. G. Atherly, eds.), Academic Press, New York, pp. 225-246 (1981).

M. J. Duncan and D. G. Fraenkel, J. Bacteriol., 37:415–419 (1979).

L. K. Dunican and F. Cannon, Plant and Soil Spec. Vol., pp. 73–79 (1971).

L. K. Dunican, R. O'Gara, and A. B. Tierney, in: Symbiotic Nitrogen Fixation in Plants (P. S. Nutman, ed.), Cambridge Univ. Press, London, pp. 77–90 (1976).

M. M. Eckhardt, I. L. Baldwin, and E. B. Fred, J. Bacteriol., 21: 273–285 (1931).

G. H. Elkan, in: Biology of the Rhizobiaceae (K. L. Giles and A. G. Atherly, eds.), Academic Press, New York, pp. 1–14 (1981).

G. H. Elkan and L. D. Kuykendall, in: Ecology of Nitrogen Fixation (W. J. Broughton, ed.), Oxford Univ. Press, London, pp. 145–166 (1981).

G. H. Elkan and I. Kwik, J. Appl. Bacteriol., 31:399–404 (1968).

H. J. Evans, D. W. Emerick, T. Ruiz-Argueso, R. J. Maier, and S. L. Albrecht, in: Nitrogen Fixation, Vol. II (W. E. Newton and W. H. Orme-Johnston, eds.), University Park Press, Baltimore, pp. 69–86 (1980).

P. F. Fottrell and P. Mooney, J. Gen. Microbiol., 59:211–214 (1969).

G. E. Fox, E. Stackebrandt, R. G. Hespell, J. Gibson, J. Maniloff, T. A. Dyer, R. W. Wolfe, W. E. Balch, R. S. Tanner, L. J. Magrum, L. B. Zablen, R. Blakemore, R. Gupta, L. Bonen, B. J. Lewis, D. A. Stahl, K. R. Leuhrsen, K. N. Chen, and C. R. Woese, Science, 209:457–463 (1980).

E. B. Fred, Va. Agric. Expt. Sta. Annu. Rept., pp. 145–173 (1911–12).

E. B. Fred, I. L. Baldwin, and E. McCoy, Root Nodule Bacteria and Leguminous Plants, Univ. of Wisconsin Press, Madison (1932).

Y. D. Gaur and H. Mareckova, Folia Microbiol., 22:323–340 (1977).

C. E. Georgi and J. M. Ettinger, J. Bacteriol., 41:323–340 (1941).

A. M. Gibbons and K. F. Gregory, J. Bacteriol., 111:129–141 (1972).

P. H. Graham, J. Gen. Microbiol., 35:511–517 (1964a).

P. H. Graham, Antonie van Leeuwenhoek J. Microbiol. Serol., 30: 68–72 (1964b).

P. H. Graham, in: Analytical Serology of Microorganisms (J. B. Kwapinski, ed.), Vol. 2, Wiley, New York, pp. 353–378 (1969).

P. H. Graham, in: Symbiotic Nitrogen Fixation in Plants (P. S. Nutman, ed.), Cambridge Univ. Press, London, pp. 99–112 (1976).

K. G. Gupta and A. N. Sen, Indian J. Sci., 35:39–42 (1965).

S. Hagashi, J. Gen. Appl. Microbiol., 13:391–403 (1967).

R. W. F. Hardy and V. D. Havelka, in: Symbiotic Nitrogen Fixation in Plants (P. S. Nutman, ed.), Cambridge Univ. Press, Cambridge, pp. 421–442 (1976).

S. Higashi, J. Gen. Appl. Microbiol., 13:391–403 (1967).

A. B. Hollis, W. E. Kloos, and G. H. Elkan, J. Gen. Microbiol., 123:215–222 (1981).

J. E. Hornez, B. Courtois, and J. Deriuex, C. R. Acad. Sci. (Paris), 283:1559–1562 (1976).

B. D. W. Jarvis, A. G. Dick, and R. M. Greenwood, Int. J. Syst. Bacteriol., 30:42–52 (1980).

A. W. B. Johnston, J. L. Beynon, A. V. Buchanan-Wollaston, S. M. Setchell, P. R. Hirsch, and J. E. Beringer, Nature (London), 276:635-636 (1978).

G. V. Johnston, H. J. Evans, and T. M. Ching, Plant Physiol., 41: 1330-1336 (1966).

D. C. Jordan, J. Bacteriol., 65:220-221 (1953).

D. C. Jordan, Can. J. Microbiol., 1:743-748 (1955).

D. C. Jordan, Can. J. Microbiol., 15:131-139 (1959).

D. C. Jordan, Bacteriol. Rev., 26:119-141 (1962).

D. C. Jordan and C. L. San Clemente, Can. J. Microbiol., 1:659-667 (1955).

H. Katznelson, Nature (London), 175:551-552 (1955).

H. Katznelson and A. C. Zagallo, Can. J. Microbiol., 3:879-884 (1957).

B. B. Keele, P. B. Hamilton, and G. H. Elkan, J. Bacteriol., 97: 1184-1191 (1969).

B. B. Keele, P. B. Hamilton, and G. H. Elkan, J. Bacteriol., 101: 698-704 (1970).

B. B. Keele, R. W. Wheat, and G. H. Elkan, J. Gen. Appl. Microbiol., 20:187-196 (1974).

G. E. Klein, P. Jemison, R. A. Haaka, and A. G. Matthysee, Experientia, 31:532-533 (1975).

A. Kondorosi and A. W. B. Johnston, in: Biology of the Rhizobiaceae (K. L. Giles and A. G. Atherly, ed.), Academic Press, New York, pp. 191-224 (1981).

A. Kondorosi, G. B. Kiss, T. Forrai, E. Vincze, and Z. Banfalvi, Nature (London), 268:525-527 (1977).

M. Kowalski, Acta Microbiol. Pol. Ser. A., 2:109-114 (1970).

L. D. Kuykendall, Appl. Environ. Microbiol., 37:862-866 (1979).

L. D. Kuykendall, in: Biology of the Rhizobiaceae (K. L. Giles and A. G. Atherby, eds.), Academic Press, New York, pp. 299-309 (1981).

L. D. Kuykendall and G. H. Elkan, Appl. Environ. Microbiol., 32: 511-519 (1976).

L. D. Kuykendall and G. H. Elkan, J. Gen. Microbiol., 98:291-295 (1977).

T. A. Lie and R. Winarno, Plant Soil, 51:135-142 (1979).

J. M. Ligon, M. Scholla, and G. H. Elkan, Proc. Am. Rhizobium Conf. (in press).

F. Löhnis and R. Hansen, J. Agric. Res., 20:543-546.

R. A. Ludwig, J. Bacteriol., 135:114-122 (1978).

S. P. Magu and A. N. Sen, Arch. Mikrobiol., 68:355-361 (1969).

R. J. Maier and W. J. Brill, J. Bacteriol., 127:763-769 (1976).

R. J. Maier and W. J. Brill, J. Bacteriol., 133:1295-1299 (1978).

L. 't Mannetje, Antonie Van Leeuwenhoek J. Microbiol. Serol., 33: 477-491 (1967).

G. Martinez de Drets and A. Arias, J. Bacteriol., 103:07-103 (1970).

G. Martinez de Drets and A. Arias, J. Bacteriol., 109:467-470 (1972).

G. Martinez-Drets, A. Gardiol, and A. Arias, Can. J. Microbiol., 20:
    605–609 (1977).

H. M. Meade and E. R. Signer, Proc. Natl. Acad. Sci. U.S.A., 74:2076–
    2078 (1977).

M. C. Meyer and S. G. Pueppke, Can. J. Microbiol., 26:606–612 (1980).

M. L. Moffett and R. R. Colwell, J. Gen. Microbiol., 51:245–266
    (1968).

K. Mulongoy and G. H. Elkan, J. Bacteriol., 131:179–187 (1977).

K. Mulongoy and G. H. Elkan, Can. J. Microbiol., 23:1293–1298 (1977).

K. Mulongoy and G. H. Elkan, Curr. Microbiol., 1:335–340 (1978).

C. Napoli and P. Albersheim, J. Bacteriol., 141:1454–1456 (1980).

C. Niel, J. B. Guillaume, and M. Bechet, Can. J. Microbiol., 23:
    1178–1181 (1977).

D. O. Norris, Plant Soil, 22:143–166 (1965).

M. P. Nuti, A. M. Lebeboer, A. A. Lepidi, and R. A. Schilperoot,
    J. Gen. Microbiol., 100:241–248 (1977).

M. P. Nuti, A. A. Lepidi, R. K. Prakash, R. A. Schilperoot, and
    F. C. Cannon, Nature (London), 282:533–535 (1979).

F. O'Gara and K. T. Shanmugan, Proc. Natl. Acad. Sci. U.S.A., 75:
    2343–2347 (1978).

J. Olivares, E. Montoya, and A. Palomares, in: Recent Developments
    in Nitrogen Fixation (W. E. Newton, R. Postgate, and C. Rod-
    riguez-Barreuco, eds.), Academic Press, New York, pp. 375–385
    (1978).

A. N. Parijskaya, Microbiologiya, 42:119–121 (1973).

F. O. Pedrosa and G. T. Zancan, J. Bacteriol., 119:336–337 (1974).

A. N. Petrova, I. E. Chermenskaya, and V. L. Kretovich, Dokl.
    Biochem., Dokl. Akad. Nauk SSSR, 217:479–480 (1974).

R. H. Prakash, P. J. J. Hooykaas, A. M. Ledeboer, J. W. Kijne, R. A.
    Schilperoot, M. P. Nuti, A. A. Ledidi, F. Casse, C. Boucher,
    J. S. Julliot, and J. Denarie, in: Nitrogen Fixation, Vol. 2
    (L. W. E. Newton and W. H. Orme-Johnson, eds.), Univ. Park
    Press, Baltimore, pp. 139–163 (1980).

S. B. Primrose and C. W. Ronson, J. Bacteriol., 141:1109–1114 (1980).

M. H. Proctor, New Zealand J. Sci., 6:17–26 (1963).

J. Rigaud and J. Trichant, Physiol. Plant., 28:160–165 (1973).

G. P. Roberts, W. T. Leps, L. E. Silver, and W. J. Brill, Appl. En-
    viron. Microbiol., 39:414–422 (1980).

C. W. Ronson and S. B. Primrose, J. Bacteriol., 139:1075–1078 (1979a).

C. W. Ronson and S. B. Primrose, J. Gen. Microbiol., 112:77–88
    (1979b).

G. B. Ruvkun and F. M. Ausubel, Proc. Natl. Acad. Sci. U.S.A., 11:
    191–195 (1980).

R. E. Sanders, R. W. Carlson, and P. Albersheim, Nature (London),
    271:240–242 (1978).

W. B. Sarles and J. Reid, J. Bacteriol., 30:651–653 (1935).

T. V. Shmyreva and T. B. Plaksina, Prikl. Biochem. Mikrobiol., 8:
    26–29 (1972).

K. A. Siddiqui and A. K. Banergee, Folia Microbiol., 20:412–417
    (1975).

W. D. Sutton, Biochim. Biophys. Acta, 366:1-10 (1974).

H. V. Tichy and W. Lotz, FEMS Microbiol. Lett., 10:203-207 (1981).

J. Trinchant and J. Rigaud, Physiol.Plant., 32:394-399 (1974).

G. Tshitenge, N. Luyindula, P. F. Lurquin, and L. Ledoux, Biochim. Biophys. Acta, 414:357-361 (1975).

K. Tuzimura and H. Meguro, J. Biochem., 47:391-397 (1960).

D. S. Ucker and E. R. Signer, J. Bacteriol., 136:1197-1200 (1978).

R. G. Upchurch and G. H. Elkan, Can. J. Microbiol., 23:1118-1122 (1977).

R. G. Upchurch and G. H. Elkan, J. Gen. Microbiol., 104:219-225 (1978).

H. J. Van Rensburg, B. W. Strijdom, and C. J. Rabie, S. Afr. J. Agric. Sci., 11:623-626 (1968).

J. M. Vincent, in: A Treatise on Dinitrogen Fixation, Vol. 3 (R. W. F. Hardy and W. S. Silver, eds.), Wiley, New York, pp. 277-366 (1977).

P. W. Wilson, J. Bacteriol., 35:601-623 (1937).

H. Zipfel, Zentr. Bakt. Parasitenk. II., 32:97-137 (1911).

W. Zurkowski, M. Hoffman, and Z. Lorkiewica, Acta Microbiol. Pol. Ser. A., 5:55-60 (1973).

W. Zurkowski and Z. Lorkiewicz, J. Bacteriol., 128:481-484 (1976).

W. Zurkowski and Z. Lorkiewicz, Genet. Res., 32:311-314 (1978).

ECOLOGY OF Rhizobium

Martin Alexander

Laboratory of Soil Microbiology
Department of Agronomy
Cornell University
Ithaca, New York 14853. U.S.A.

The inability to predict why legumes develop poorly and fix little nitrogen in fertile soil containing Rhizobium strains active in the symbiosis frequently results from the lack of knowledge of the ecology of the root-nodule bacteria. If the appropriate host and an effective bacterium are both present in fertile land, there ought to be active nitrogen fixation. The fact that often no such appreciable nitrogen gain is evident or the fixation is insufficient to meet the crop demand for this element is not a function of a genetically poor pair of symbionts under these conditions, but rather is a consequence of the inability of the bacteria to survive, grow in soil, colonize the roots, or bring about nodule formation.

The ideal strain of Rhizobium is one that is able to survive well in the soil, to grow at the opportune moment, to colonize the roots readily and in the presence of organisms that are antagonistic to it, and then to initiate the nitrogen-fixing symbiosis. Similarly, the ideal strain for seed inoculation has these properties but also possesses the capacity to survive in large numbers on seeds so that it becomes established once the seed is planted. The paucity of knowledge of factors affecting the ecology of inoculum strains is reflected in the frequent failure of inoculated legumes to develop active nitrogen-fixing symbioses.

The subject of Rhizobium ecology has attracted a number of investigators, and their studies have been summarized in several recent reviews (Alexander, 1977; Lowendorf, 1980; Nutman et al., 1978; Schmidt, 1978). Because of the existence of these reviews, certain aspects of the ecology of Rhizobium will not be considered in this review.

The oldest method for assessing the numbers of rhizobia in soil is the most-probable-number technique. In this procedure, the soil sample is diluted, and portions of the dilutions are applied to plants. Those plants that develop nodules are scored as positive, and the counts are then made using appropriate most-probable-number tables together with the dilution used to obtain the final positives (Wilson, 1926). The original method has been modified several times, but the essential details are the same. The procedure, however, is not popular because of the amount of work involved and the need for facilities for growing many plants.

A method that has been used by several investigators relies on immunofluoroescent methods and involves observing certain of the antigenic characteristics of the rhizobia. The method is quite suitable provided that the population in soil is reasonably large and the investigator has equipment for immunofluorescence (Bohlool and Schmidt, 1970). A simple technique that does not relay on microscopic procedures but requires the availability of antibiotics and antibiotic-resistant strains of Rhizobium involves the use of genetically marked isolates. The isolates used are resistant to antibiotics in concentrations that few other bacteria or fungi can grow in agar media containing the inhibitors. If such isolates are obtained, they behave like the parent strains, and they do not readily revert to the antibiotic-sensitivity characters, then populations containing few cells per unit weight of soil can be counted. These two methods give quite similar results, and they appear to be suitable for identifying strains of Rhizobium in the field and for their enumeration (Brockwell et al., 1977).

The investigator of Rhizobium ecology must bear in mind that different species of Rhizobium, and indeed different strains, behave differently in soil. This is evident from the results of many earlier as well as recent studies of populations in the field, in soil in the laboratory, and in greenhouse studies of the rhizosphere. Thus, some rhizobia are numerous in soil while others are rare; some die out readily under stress conditions, whereas others survive for long periods; some readily colonize the roots of legumes and non-legumes, whereas others are only slowly able to colonize the rhizosphere and fail to cope with antagonists in the rhizosphere. Unfortunately, few cultural characteristics are presently known by which it is possible to distinguish among the species and strains in regard to their behavior in soil and in the root zone.

Numerous assessments have been made of the numbers of these bacteria under field conditions. The available evidence indicates that their populations may be fewer than 10 but occasionally may exceed 10^6 per gram. Despite these many studies, the more recent of which often involve assessments based on antigenic characteristics of the isolates, it is not yet possible to predict the behavior of any given strain. Such studies have shown a marked effect of season

of year in the temperate zone; thus, the numbers tend to decline as
the winter approaches, and the rhizobia appear to become more numerous as the temperature rises in the spring (Wilson, 1930). However,
even these few seasonal studies have been largely restricted to several soils in the temperate zone.

Early investigators noted that populations of the rhizobia were
generally larger in fields that had a rotation containing the homologous legume than in land supporting a rotation in which the legume
was absent, and they observed that the addition of plant remains and
manure enhanced the development of these bacteria (Walker and Brown,
1935). On the other hand, more recent studies found no correlation
in the numbers of R. japonicum and the frequency of growing soybeans
or the numbers of years since the land supported this legume (Weaver
et al., 1972).

Surprisingly little attention has been paid to the growth of
rhizobia in soil or even in the rhizosphere. The existing data suggest that growth in the soil is generally infrequent, except in a
few special circumstances. However, in the presence of germinating
seeds and developing roots, the growth of rhizobia may be appreciable, and some strains reach large numbers in a few days. These
data indicate that germinating seeds and developing roots excrete
compounds that rhizobia can use, even in the presence of rapidly
growing bacteria that are actively competing with the root-nodule
organisms. Rhizobia may also grow somewhat after dry soil has been
remoistened, an enhancement in their replication that is likely associated with the release of available carbon compounds during the
wetting-drying cycle; nevertheless, the decline during the period of
drying is greater than the increase in numbers following remoisten-
ing of the dried soils. The few types of crop residues that have
been tested do not enhance growth of rhizobia after these residues
are added to the soil, although possibly other kinds of crop remains
may be stimulatory. On the other hand, the rhizobia are stimulated
markedly by additions to soil of simple, readily usable organic
compounds, such as mannitol, but much mannitol must be added to
bring about significant growth. It is likely that a large amount of
sugar must be added to allow for even a small increase in rhizobial
abundance because these slow growing bacteria are poor competitors
with the faster growing bacteria in the same habitat (Pena-Cabriales
and Alexander, 1983b).

The plant root is a unique habitat that is favorable to Rhizobium, and large numbers of individual strains are frequently found
around roots of appropriate host legumes, nonhost legumes, or even
nonleguminous species (Rovira, 1961; Pena-Cabriales and Alexander,
1983a). Under these conditions, the populations of rhizobia often
grow readily as the seed imbibes water and the roots began to develop (Ramirez and Alexander, 1980; Lennox and Alexander, 1981).

In contrast, roots of certain plants, or their exudates, are toxic
to Rhizobium (Peters and Alexander, 1966; Ranga Rao et al., 1973).
The extent of stimulation is dependent on the particular soil, the
plant species, and the individual rhizobium (Tuzimura et al., 1966).
Few long-term assessments of rhizobia associated with cropping practices have been carried out, but one interesting study indicates
that the population of R. trifolii was 10^5 to 10^6 in the presence
of clovers, about 10^4 in land planted to cereals, but only about 10
to 100 per gram in English soils in continuous fallow (Nutman and
Hearne, 1980).

Thus, Rhizobium rarely grows in soil. The period of proliferation is probably chiefly associated with the presence of roots that
enhance bacterial development, and the root stimulation is apparently not specific for the homologous microsymbiont. Following
this period of growth in the rhizosphere, and possibly the period
of replication associated with crop remains that are decaying in
soil, the rhizobia are then not able to grow, and a major ecological
issue is how well they survive in the absence of growth. For this
reason, information on factors affecting survival is especially
significant.

Survival is easily measured using antibiotic-resistant isolates.
In some of the earlier studies, streptomycin-resistant mutants were
used, and the numbers were determined on an agar medium containing
this antibiotic at concentrations that few bacteria derived from
soil would tolerate. The fungi native to the soil were inhibited
by addition to the agar of a chemical toxic to the fungi but having
no effect on bacteria. Information from these studies indicates
that, if the numbers of rhizobia added to soil are not overly large,
their numbers do not fall markedly when the organisms are introduced
into soils that are moist and are at temperatures not at the high or
low extremes (Danso and Alexander, 1974). If small populations of
rhizobia are introduced into the soil, their decline is also slow
(Liang and Alexander, unpublished data).

Considerable research has been conducted to determine some of
the stress factors affecting the survival of rhizobia. One of the
chief factors is temperature. In many tropical soils, the surface
temperature reaches and occasionally exceeds 40°C. Under these
conditions, the numbers often decline readily (Iswaran et al., 1970).
Some rhizobia may be adapted to low temperatures, so that legumes
grown in northern temperate climates may benefit if the rhizobia
added at planting time are uniquely able to colonize the roots and
bring about nodulation at low temperatures (Ek-Jander and Fahraeus,
1971).

It may be expected that the nodule is a major source of the
rhizobia in soil inasmuch as this plant organ contains a very dense
population. However, early studies of Wilson (1934) showed that the

population of R. japonicum derived from the nodule, although initially large, is quite transitory and does not persist in the soil;
the population declined to levels that were not detected by the
methods then used. Several years after the growth of the soybeans,
the numbers of R. japonicum were not greater than in soil not cropped
to soybeans. Under controlled conditions, the population of R.
japonicum derived from a nodule homogenate declined from 10^8 to 1
million per gram of soil in 30 days (Pena-Cabriales and Alexander,
1983).

Rhizobium movement away from a source of the bacteria has also
been evaluated. R. trifolii is unable to move significant distances.
For example, the population 3.0 cm from the site of the original
population is several orders of magnitude less than that added, and
sometimes the rhizobia can not be detected 3.0 cm away (Hamdi, 1971).
A lack of vertical translocation has been noted for R. japonicum
(Madsen and Alexander, 1982). The studies of Hamdi (1974) also suggested that precipitation might influence movement, but recent work
suggests that percolating water, as well as earthworms alone, have
only a small effect on the vertical dissemination of R. japonicum,
although its movement is enhanced if percolating water is combined
with an earthworm or developing roots. In no instance did the bacterium move in significant numbers (Madsen and Alexander, 1982).
Field studies also indicate a minimum degree of lateral movement
(Chatel and Greenwood, 1973).

The drying of soil is a major stress. Early studies involving
the most-probable-number technique revealed a marked decline when
soils containing three different rhizobia were dried, although the
rate and extent of population reduction varied with the soil and
the test bacterium (Foulds, 1971). Nevertheless, it was early recognized that at least a few cells remain viable for periods in excess
of 30 years (Jensen, 1961). Bushby and Marshall (1977a, 1977b)
evaluated population changes in sterile soils that were inoculated
and then rapidly dried. Their data show a very rapid die-off under
the artificial conditions of rapid drying and also point to differences among species, or possibly strains, in susceptibility to drying. The survival of some rhizobia was enhanced if montmorillonite
or certain organic materials were added to protect against desiccation in the test soil (Bushby and Marshall, 1977a). They proposed
that the differences in susceptibility among rhizobia were related
to the amount of water retained by the cells during the process of
drying.

Somewhat different results are obtained if the drying conditions in the laboratory are like those in nature. Under such circumstances, the decline in numbers is initially rapid as the soil
quickly loses moisture, but the subsequent rate of decline is more
slow, coinciding with the slow subsequent loss of water from the
soil. In tests of several strains and species of Rhizobium, the

population decline is about two logarithmic orders of magnitude in
one drying cycle. However, when the bacteria are exposed to repeated
cycles of wetting and drying, the population is reduced at each drying cycle; based on such data, it seems likely that the survivors of
the first drying cycle are not intrinsically drought resistant because they die when the soil is moistened and dried once again. Decomposed organic materials did not afford protection against drying
(Pena-Cabrioles and Alexander, 1979).

The percentage of cells of cowpea rhizobia and R. japonicum
surviving is directly related to the quantity of water remaining
following desiccation and is also influenced by relative humidity
(Osa-Afiana and Alexander, 1982a). The characteristics of soils
that are associated with the differences in survival during drying
have also been tested, and it has been observed that the survival
of R. meliloti and R. phaseoli is poor in soils more acid than pH
5.7 and that the number of R. phaseoli decreases with increasing
clay content. Moreover, the abundance of survivors is directly
related to the organic matter content for soils having less than 2
but not for soils having more than 3% organic matter (Chao and Alexander, 1982).

Some investigators assume that strains of a single species or
cross-inoculation group behave quite similarly in soil. That this
is not true is evident in a study of the survival of cowpea rhizobia
in soil. Thus, in a single drying cycle, only 0.13% of some strains
survived desiccation, whereas 50% of the cells of other strains en-
dured following a single drying step (Osa-Afiana and Alexander,
1982b).

Soils in the tropics are frequently exposed simultaneously to
high temperature and drying; hence, investigations have been conducted to determine the simultaneous impact of these two stress factors. In part, this concern arose because of a problem noted with
clovers grown in certain sandy soils of Australia. In these soils,
the plants became established and nodulated in the first season, but
nodulation did not occur and the plants failed to survive in the
second season. Marshall et al. (1963) concluded that the absence of
nodulation in the second season and the concomitant mortaility of
clover resulted from the inability of the Rhizobium to persist from
one season to another. In these studies, it was found that some
strains of R. trifolii survived in some but not in other soils that
reached 70°C and that the harmful effect of high temperatures could
be reduced by addition to the soil of some but not other clays
(Marshall, 1964). Interestingly, high populations of R. lupini were
maintained during the hot, dry period at times when the population
of R. trifolii declined (Chatel and Parker, 1973). These Australian
investigations are excellent examples of the relationship between
ecological knowledge of the bacteria and practical problems of
legume growth.

Although soil freezing is a repeated stress in soils of the temperate zone, few studies have been devoted to its impact. Nevertheless, some rhizobia withstand freezing during the winter and initiate nodulation the following spring (Fred, 1920).

Soil moisture and water potential have a major impact on the numbers of survivors. The ability to tolerate moisture stress varies with the species and strain of Rhizobium. In soils that are puddled or are flooded for some time, the decline is especially marked, and under these circumstances, organic acids may be involved in death of the rhizobia (Mahler and Wollum, 1980; Osa-Afiana and Alexander, 1979).

The importance of soil pH has long been recognized. In the early studies of Bryan (1923), it was shown that individual strains of Rhizobium varied in their tolerance of soil pH, some dying at pH 5.0, others tolerating pH 4.5, and the test strain of R. japonicum surviving to somewhat below pH 4.0. Because of the detrimental effect of acidity, it is not surprising that addition of $CaCO_3$ enhanced survival (Thornton, 1940). In more recent work, the population changes linked with soil acidity have been carefully monitored. For example, it has been noted that a cowpea isolate is not greatly affected in soils of pH 4.4 in the absence of microbial antagonism, although its population fell slowly at pH 4.1. In contrast, a strain of R. phaseoli survives well in soils of pH 4.4 or higher but declines in a sterile soil at pH 4.3. The data also indicate that differences in susceptibilities of strains of R. meliloti can be exploited and that a strain of greater acid tolerance endures in a highly acid soil (Lowendorf et al., 1981).

As pointed out above, Rhizobium rarely reaches large numbers in soil. However, such large populations are found during the initial phases of plant growth as the roots are colonized, as nodules decay and release the bacteria contained within, and following the addition of commercial inoculants to seeds sown in the soil. After such large numbers appear, a rapid die-off is soon evident. The agents of the decline do not appear to be bacteria or other microorganisms competing with root-nodule bacteria, nor do they appear to be species producing toxins or lytic enzymes. The agents causing the death of many of the cells appear to be members of the protozoan community. Thus, in moist, nonacid soils at moderate temperatures, no significant decline in rhizobial count occurs following their introduction into sterile soil, but the cell density falls abruptly in nonsterile soils. Coinciding with the diminution in the rhizobial numbers is an increase in the abundance of protozoa (Danso et al., 1975). However, predation by the protozoa does not eliminate the rhizobia but rather leaves a reasonaly large population. The inability of the protozoa to eliminate all of these bacteria is not a result of other organisms attacking the protozoa, nor is it a consequence of the presence in soils of microenvironments in which the

surviving rhizobia are protected from the marauders (Danso and
Alexander, 1975; Habte and Alexander, 1978a). More direct evidence for a role of protozoa in the decline was obtained by sup-
pressing protozoan activity by using an inhibitor of these predators;
thus, the number of rhizobia do not fall as long as the protozoa
native to the soil are suppressed by the inhibitor, but once in-
hibitor-tolerant protozoa appear, Rhizobium declines in its usual
fashion (Habte and Alexander, 1977). Based on such findings, it
seems clear that protozoa cause the reduction in the artificially
high numbers of Rhizobium that occasionally exist.

To explain why so many Rhizobium still endure despite the co-
existence of many protozoa, it has been hypothesized that the
predator is unable to feed actively when the energy used in hunting
for the few remaining Rhizobium cells is equal to that obtained by
the protozoa in their feeding. At the same time, Rhizobium is
slowly multiplying to compensate for the few cells of this genus
that are being eliminated by the continued, but slow, grazing by
the protozoa (Danso et al., 1975; Habte and Alexander, 1978b).

Bacteriophages are present in many soils, and often these
viruses are found in soils supporting nodulated legumes (Katznelson
and Wilson, 1941). Nevertheless, and despite the many suggestions,
particularly in the early literature, that bacteriophages are important in reducing the amount of nitrogen fixed or the vigor of
the legumes, no substantive evidence exists that these viruses are
important in nature, nor that they influence the ecology of the
rhizobia. Indeed, based on current views of density dependence of
predators and parasites, it is unlikely that bacteriophages would
have a significant impact on the ecology of the root-nodule bacteria, except when the population of an individual strain becomes
quite high (Alexander, 1981); even here, however, it is protozoa
and not viruses that appear to be significant.

Bdellovibrio is also present in many soils. These minute bacteria can attack the rhizobia in liquid medium and bring about a
marked reduction in their numbers. However, even in culture, many
rhizobia must be present for Bdellovibrio to initiate replication,
population densities of rhizobia far greater than those usually
found in nature. Moreover, the characteristic decline of large
populations of Rhizobium, a decline that is paralleled by the expected increase in numbers of protozoa, does not coincide with the
expected increase in abundance of the vibrios. Hence, it does not
seem likely that even the initial fall in the occasionally large
numbers of rhizobia in soil can be attributed to Bdellovibrio (Keya
and Alexander, 1975).

Some evidence exists that soils contain inhibitors affecting
the root-nodule bacteria. Thus, aqueous extracts of certain soils
that do not support nodulation of subterranean clover contain an

inhibitor of R. trifolii, and it has been proposed that toxin-pro-
ducing microorganisms, which proliferate at the expense of the organic matter remaining from the original vegetation, generate inhibitors affecting the Rhizobium (Holland and Parker, 1966). These
toxic soil extracts affect R. trifolii but not R. lupini (Chatel and
Parker, 1972). Although the view that inhibitors in soil suppress
the development of rhizobia is attractive, strong evidence in its
support is not presently available.

Fungicides applied to seeds represent a severe stress. Such
pesticides are designed to inhibit the activity of pathogenic fungi
that affect the seed or seedling. At the same time, the chemicals
kill rhizobia. Thus, while a benefit is being obtained by controlling pathogens, harm is being done because the plant that appears
will have few nodules derived from the active nitrogen-fixing strains
used in inoculants. Several means have been devised for overcoming
the deleterious effects of fungicides, including modifying placement
of the chemical vs. the rhizobia and using fungicides with lower
toxicity to the root-nodule bacteria. Still another approach is to
use fungicide-resistant Rhizobium for seed inoculation. By such
means, pathogens can be suppressed or controlled, and the rhizobia,
because of their innate resistance, are unaffected. This procedure
allows nodulated plants to appear with no detectable detrimental
effects on nodulation (Odeyemi and Alexander, 1977). However, these
studies led to a surprising finding: the fungicide, although designed to inhibit fungi, also suppressed protozoa proliferating in
the rhizosphere and using the large numbers of rhizosphere bacteria,
including rhizobia, as prey. Thus, the fungicide-effected suppression of protozoa is reflected in an improved colonization of fungicide-resistant R. phaseoli. The data suggest that protozoa are important in reducing rhizobial numbers in the rhizosphere, a zone in
which the rhizobial population becomes large as seeds and developing
roots excrete nutrients used by the root-nodule bacteria (Ramirez
and Alexander, 1980; Lennox and Alexander, 1981).

A picture is thus emerging of the ecology of Rhizobium. The
organisms are not particularly good competitors, and they rarely,
if ever, grow in circumstances typical of normal farming practices.
The bacteria proliferate chiefly or solely in the presence of germinating seeds and developing roots, and thereafter, the population
is subjected to a variety of biological and nonbiological stresses.
At high population densities, protozoa appear to be significant in
killing rhizobia, but the biological mechanism causing the decline
of small populations is unknowm. The organism is quite susceptible
to extremes of acidity and temperature, and its numbers are greatly
reduced when soils are dryed. Pesticides may also reduce their numbers.

This ecological information can be exploited practically. For
example, strains have long been known to differ appreciably in ni-

trogen-fixing ability, and only the active strains are used in inoculants. Similarly, the more "competitive" strains are commonly used in inoculants; however, the word competition may be being misused in this context because it is not clear if the more "competitive" strain is really competing with rhizobia, with other bacteria, or for nodulation sites, or if it is merely a better colonizer of the rhizosphere in the presence of unkown biological and abiotic stresses. In addition, because strains differ appreciably in their tolerance to major ecological stresses (e.g., desiccation, temperature, and acidity), strains tolerant of such abiotic stresses should be used in preparing inoculants so the added rhizobia are better able to endure in soil following inoculation. Moreover, desiccation-resistant strains should be used for seed inoculation, but the major stress of drying on seeds has been ignored so far in selecting bacteria appropriate for inoculants. Furthermore, because Rhizobium is subject to protozoan predation as the root system is developing - this being the time when the rhizobia must be abundant because it is the period when rhizobia infect the roots prior to nodulation - inhibition of protozoa, by fungicides or more selective antiprotozoan chemicals, may be a practical means to promote nodulation and subsequent nitrogen fixation.

Ecology of Rhizobium is a developing field. The information will facilitate understanding the behavior of these organisms in soil and will help to predict some of the interactions of the bacteria with their leguminous hosts. Furthermore, the information can be put to practical use and help improve food production. It is not always the case that microbiologists and soil scientists can combine their interests in basic science and in food production, but this field appears to offer a wonderful example of such a partnership.

REFERENCES

M. Alexander, in: Biological Nitrogen Fixation in Farming Systems
    of the Tropics (A. Ayanaba and P. J. Dart, eds.), Wiley,
    Chichester, U.K., pp. 99-114 (1977).
M. Alexander, Annu. Rev. Microbiol., 35:113-133 (1981).
B. B. Bohlool, and E. L. Schmidt, Soil Sci., 110:229-236 (1970).
J. Brockwell, E. A. Schwinghamer, and R. R. Gault, Soil Biol. Bio-
    chem., 9:19-24 (1977).
O. C. Bryan, Soil Sci., 15:37-40 (1923).
H. V. A. Bushby and K. C. Marshall, Soil Biol. Biochem., 9:143-147
    (1977a).
H. V. A. Bushby and K. C. Marshall, J. Gen. Microbiol., 99:19-27
    (1977b).
W.-L. Chao and M. Alexander, Soil Sci. Soc. Am. J., 46:949-952
    (1982).
D. L. Chatel and R. M. Greenwood, Soil Biol. Biochem., 5:799-808
    (1973).

S. K. A. Danso and M. Alexander, Soil Sci. Soc. Am. Proc., 38:86–89
    (1974).
D. L. Chatel and C. A. Parker, Soil Biol. Biochem., 4:289–294
    (1972).
D. L. Chatel and C. A. Parker, Soil Biol. Biochem., 5:415–423
    (1973).
S. K. Danso and M. Alexander, Appl. Microbiol., 29:515–521 (1975).
S. K. A. Danso, S. O. Keya, and M. Alexander, Can. J. Microbiol.,
    21:884–895 (1975).
J. Ek-Jander and G. Fahraeus, Plant Soil (Spec. Vol.), pp. 129–137
    (1971).
E. B. Fred, Wisconsin Agric. Exp. Sta. Bull., 323, Madison, Wis-
    consin (1920).
W. Foulds, Plant Soil, 35:665–667 (1971).
M. Habte and M. Alexander, Arch. Microbiol., 113:181–183 (1977).
M. Habte and M. Alexander, Ecology, 59:140–146 (1978a).
M. Habte and M. Alexander, Soil Biol. Biochem., 10:1–6 (1978b).
Y. A. Hamdi, Soil Biol. Biochem., 3:121–126 (1971).
Y. A. Hamdi, Zentr. Bakteriol. Parasitenk., II, 129:373–377 (1974).
A. A. Holland and C. A. Parker, Plant Soil, 25:329–340 (1966).
V. Iswaran, W. V. B. Sundara Rao, K. S. Jauhri, and S. P. Magre,
    Mysore J. Agric. Sci., 4:105–107 (1970).
H. L. Jensen, Nature (London), 192:682–683 (1961).
H. Katznelson and J. K. Wilson, Soil Sci., 51:59–63 (1941).
S. O. Keya and M. Alexander, Soil Biol. Biochem., 7:231–237 (1975).
L. B. Lennox and M. Alexander, Appl. Environ. Microbiol., 41:404–
    411 (1981).
H. S. Lowendorf, A. M. Baya, and M. Alexander, Appl. Environ. Micro-
    biol., 42:951–957 (1981).
E. L. Madsen and M. Alexander, Soil Sci. Soc. Am. J., 46:557–560
    (1982).
H. S. Lowendorf, A. M. Baya, and M. Alexander, Appl. Environ. Micro-
    biol., 42:951–957 (1981).
R. L. Mahler and A. G. Wollum, II, Soil Sci. Soc. Am. J., 44:988–
    992 (1980).
K. C. Marshall, Austr. J. Agric. Res., 15:273–281 (1964).
K. C. Marshall, M. J. Mulcahy, and M. S. Chowdhury, J. Austral.
    Inst. Agric. Soc., 29:160–164 (1963).
P. S. Nutman, M. Dye, and P. E. Davis, in: Microbial Ecology,
    Springer-Verlag, Berlin, pp. 404–410 (1978).
P. S. Nutman and R. Hearne, Report, Rothamsted Exp. Sta., Part 2,
    Harpenden, Herts., U.K., pp. 7–90 (1980).
O. Odeyemi and M. Alexander, Soil Biol. Biochem., 9:247–251 (1977).
L. O. Osa-Afiana and M. Alexander, Soil Sci. Soc. Am. J., 43:925–
    930 (1979).
L. O. Osa-Afiana and M. Alexander, Soil Sci. Soc. Am. J., 46:285–
    288 (1982a).
L. O. Osa-Afiana and M. Alexander, Appl. Environ. Microbiol., 43:
    435–439 (1982b).

J. J. Pena-Cabriales and M. Alexander, Soil Sci. Soc. Am. J., 43:
      962-966 (1979).
J. J. Pena-Cabriales and M. Alexander, Soil Sci. Soc. Am. J., 47:81-
      84 (1983a).
J. J. Pena-Cabriales and M. Alexander, Soil Sci. Soc. Am. J., 47:
      241-245 (1983b).
R. J. Peters and M. Alexander, Soil Sci., 102:380-387 (1966).
C. Ramirez and M. Alexander, Appl. Environ. Microbiol., 40:492-
      499 (1980).
V. Ranga Rao, N. S. Subba Rao, and K. G. Mukerji, Plant Soil., 39:
      449-452 (1973).
A. D. Rovira, Austr. J. Agric. Res., 12:77-83 (1961).
E. L. Schmidt, in: Interactions between Non-Pathogenic Soil Micro-
      organisms and Plants (Y. R. Dommergues and S. V. Krupa, eds.),
      Elsevier Scientific Publishing Co., Amsterdam, pp. 269-303
      (1978).
G. D. Thornton, Soil Sci. Soc. Am. Proc., 8:238-240 (1943).
K. Tuzimura, I. Watanabe, and J. F. Shi, Soil Sci. Plant Nutr., 12:
      99-106 (1966).
R. H. Walker and P. E. Brown, J. Am. Soc. Agron., 27:289-296 (1935).
R. W. Weaver, L. R. Frederick, and L. C. Dumenil, Soil Sci., 114:
      137-141 (1972).
J. K. Wilson, J. Am. Soc. Agron., 18:911-919 (1926).
J. K. Wilson, Soil Sci., 30:289-296 (1930).
J. K. Wilson, New York (Cornell) Sta. Mem. 162, Ithaca, New York
      (1934).

IMPORTANT LIMITING FACTORS IN SOIL

FOR THE Rhizobium-LEGUME SYMBIOSIS

J. R. Jardim Freire

Department of Soils
Faculdade de Agronomia, UFRGS
Porto Alegre, RS, Brazil

INTRODUCTION

Success in obtaining high N_2 fixation by the Rhizobium-legume symbiosis depends on a series of interacting factors: (a) effectiveness and efficiency of the Rhizobium strains present in the inoculum and/or soil in relation to the species and varieties of the legume; (b) competitive ability of the introduced rhizobia in relation to the native rhizobial population; (c) ability of the host to supply its microsymbiont's nutritional needs; and (d) environmental factors, especially limiting factors in soil, that act on the bacteria and the host. The environment may directly affect the rhizobia or indirectly influence the bacteria through the plant. Fortunately, most of the conditions that benefit the bacteria and enhance N_2 fixation will also be beneficial to the plant, and when conditions are optimum for the nutrition and growth of the plant, the needs of the bacteria will also be supplied and N_2 fixation will be high.

Possibly the main advantage of the use of leguminous crops is the potential for obtaining free nitrogen from the air. However, this advantage will be realized only if the symbiosis of plant with bacteria can operate effectively. Inoculation of the seeds or the existence of a high population of efficient Rhizobium strains in the soil is not per se a guarantee of the formation and functioning of an adequate number of nodules. What is frequently not well understood or forgotten by farmers and even agronomists is that the legumes have two possible sources of nitrogen: soil and air. Several environmental factors acting on the bacteria, the plant, or the bacterium-plant interaction may limit or totally inhibit N_2 fixation and thus reduce yields unless the soil is rich in combined nitrogen.

51

Inhibition of nodule formation, N_2 fixation, or both will also result in a diminished response to other nutrients, a reduced yield, and a partial or total loss of the financial investment in other inputs to the cropping system.

In the developing countries of the tropics and subtropics, agricultural production is usually poor, except in the rare soils of high fertility. Most of the soils are acidic, low in phosphorus and sulfur, and often contain toxic levels of Al and Mn. Acid, infertile Oxisols and Ultisols are widespread. Millions of square kilometers in South America are covered with the "cerrado" or "savanna," which not only have deficiencies of macronutrients and toxic levels of Al and Mn but also are poor in one or more micronutrients.

In the principal crop-producing regions of Brazil, 85 to 99% of the soils contain low or very low levels of available P and 1 to 6 meq of exchangeable Al if unlimed, and generally require 3 to 12 metric tons of lime per ha to achieve a reasonable correction of undesirable soil acidity (pH 5.5 to 6.0). Soil erosion, unresponsive varieties, and poor cultural practices also limit legume growth. Part of this region that is located in southern Brazil has been made highly productive by an integrated research and extension program (Beaty et al., 1972) that started 15 years ago, and after 5 years, the recommendations for liming and fertilization had become widely accepted.

Studies of liming and fertilization in this region were conducted on the assumptions that the lime and fertilizers were expensive and that the farmer could not afford to use them. In consequence, crop responses were poor or absent. At that time the extension agencies recommended the use of 0.5 to 1 ton of lime per ha "because lime was very expensive." And despite the widespread use of inoculants for soybeans prepared from locally selected R. japonicum strains, the field results were usually poor except in a few areas where the soil was not toxic to the plants and where P deficiency had been overcome.

The philosophy of minimum-input (and low productivity) agriculture is dangerous, as it might be a way to keep the farmer poor. It is necessary to demonstrate alternatives, at least in small areas, conducive to the use of high inputs (lime, fertilizers, and responsive varieties) and the attainment of high yields as a way to show to the farmer that he can also obtain more from his land. In this way, there is an enhanced demand for lime and fertilizers, a stimulation of the local production, and lower prices. In this connection, administrators and researchers in the tropics should pay attention to the paper of Fox and Kang (1976), who examined and criticized the philosophy of minimum inputs so common in the tropics. The example mentioned above is a good one. A vicious cycle existed of high prices and low use of lime and fertilizers. In five years (1967 to 1972), the use of lime increased from 80,000 to 2 million tons. and the lime became cheaper. At the same time, the large increase in the use of P and K, together with correction of Al and Mn toxicities, changed the picture in relation to the nodulation of soybeans and the achievement of high N_2 fixation. If it is assumed that N_2 fixation now provides 50 kg N/ha/year, the contribution of the Rhizobium-soybean symbiosis is about 600 thousand tons of N for the 12 million hectares cropped to this legume in 1981.

In a previous paper (Freire, 1976), several authors were cited who review the literature on the effect of soil factors on the Rhizobium-legume symbiosis. In this paper, I shall draw some conclusions from those reviews and also from the more recent literature.

The nodulated legume is a much more complex system than the plant that depends only on the available mineral nitrogen in the soil. After recognition occurs between bacteria and plant, root hairs become infected, and nodules are formed. A close relationship is established between the partners and between them and the environment. The effectiveness and efficiency of the Rhizobium-legume symbiosis will depend on these relationships.

The factors limiting the bacteria include the ability to compete, their capacity to survive, and hydrogenase activity. The factors limiting the host include rates of photosynthesis and nutrient uptake. Environmental factors include the P, K, Ca, N, micronutrient, and moisture content and temperature and pH of the soil.

PHOSPHORUS

Soils rich in available P are not common in Latin America or elsewhere in the tropics and subtropics. On the contrary, P deficiency is the most important single limiting factor for N_2 fixation and legume production. It has been shown that plants dependent on N_2 require more P than plants using mineral N (France, 1976). This need reflects the vital role of P in energy transfer and the large quantity of energy required for the reduction of N_2 to NH_3.

Phosphorus deficiency in many acid soils is complicated by adsorption by Fe and Al oxides, clay particles, and exchangeable Al and Mn. The use of P fertilizer without liming in these soils is associated with high P fixation, and such practices do not alleviate the toxic effects of Al and Mn on nodulation and N_2 fixation (Geopfert and Freire, 1972; Freire, 1976). On the other hand, in acid soils poor in P but with no toxicity problems, the importance of P to legume nutrition also becomes evident, when for example, fertilizer is applied for the improvement of native pastures. In many areas of southern Brazil, the use of only phosphorus fertilizers results in the appearance of a dense population of native legumes that were previously uncommon. In these soils, the residual effect of P is great. Scholles (1978) found a significant residual effect of P fertilization 4.5 years after its application to two tropical forage legumes.

The addition of P to soil usually causes an increase in the concentrations of P and N in plant tissues. This relationship has been shown for tropical pasture legumes (Andrew, 1976). Geopfert (1971) reported a close relationship in soybeans between grain yield, P availability in the soil, and nodule weight of soybeans (Fig. 1). Kolling et al. (1974) studied the effects of lime, P, and K on three soybean cultivars in an acid soil with mild Al toxicity (pH 4.7, 1.5 meq Al/100g, 6 ppm P, 63 ppm K) in the four year of the lime and fertilizer applications. The responses to P and K were highly significant but nodulation was not affected (Table 1).

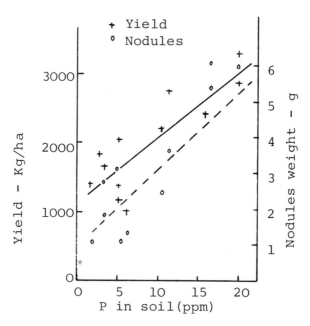

Fig. 1.   Regressions of available soil P versus grain yield and nodule dry weight of soybeans. Soil: B. Retiro, RS (Goepfert, 1971).

TABLE 1. Effect of P, K and Lime on Nodulation of Soybeans (from Kolling et al., 1974)

| Treatment[a] | Varieties | | | | | |
| | Planalto | | Hardee | | Bragg | |
	Yield[b]	Nodules[c]	Yield	Nodules	Yield	Nodules
P0	673	0.7	988	1.2	346	0.9
P1	1982	2.0	1587	2.1	1656	2.9
P2	1842	2.3	1768	2.4	2042	3.9
K0	1084	1.3	1092	1.7	1051	1.9
K1	1449	1.5	1548	2.1	1779	2.7
K2	1965	2.2	1704	1.8	1713	3.1
Ca 0	1056	1.8	1180	2.0	977	2.7
Ca 1	1599	1.5	1417	1.9	1608	2.5
Ca 2	1834	1.6	1745	1.9	1958	2.6

[a]Treatments: P1, P2, K1, and K2 denote 100, 200, 50, and 100 kg/ha/year, respectively. Ca 1 and Ca 2 denote 3 and 6 tons lime/ha in year 1.
[b]kg/ha.
[c]g/10 plants.

According to Andrew (1976), the increase in N concentration in legumes resulting from P additions may be compounded by time and extent of nodulation, and by duration and efficiency of the symbiosis. Andrew (1976) reviewed the information available on the effect of P on tropical pasture legumes and concluded that nodule initiation, number, volume, and dry weight were enhanced. In contrast, no effect was observed on total nodule weight or nitrogen fixed per unit weight of nodule on Glycine wightii. Andrew (1976) posed the question whether the effect of P additions results from a primary effect on nodule initiation and weight, improved plant nutrition and photosynthesis, or more efficient uptake of soil N. However, the close relationship between N and P concentrations in plant tops and dry matter yield suggests an unequivocal effect of P on N production by legumes; e.g., the concentration of P coinciding with maximum dry matter production is usually less than that corresponding to maximum N concentration. The upper N concentration also varies with each species. The relationship between N and P levels may be used to estimate the relative efficiency of the legume-Rhizobium symbiosis in relation to P.

Among the tropical forage legumes, some species are more efficient in using P than others. Differences also occur among genera. Legumes that respond relatively little to added P usually have relatively low critical percentages of P and they have a higher capacity for short and long term utilization of insoluble phosphates. The explanation may lie in the interactions between plant and soil, in some mycorrhizal symbiosis or both (Andrew, 1976).

Mosse (1976) believes that the effect of vesicular-arbuscular mycorrhizae in improving phosphate supply to plants results from the absorbing capacity of the extensive network of external hyphae associated with the infected roots. Nodulation of various legume species has been shown to be responsive to the inoculation with mycorrhizae. Mosse (1976) cited the work of Assai, who reported that only 4 of 59 legume species examined were not mycorrhizal, and also of Strzemska, who confirmed the widespread occurrence of mycorrhizae in species of Papilionaceae. Growth and nodulation of various species are stimulated by mycorrhizal infection, and the beneficial effect of mycorrhizae on P uptake and nodulation usually occurs in P-deficient soils. Thus, the search for efficient mycorrhizae for legumes is important for minimum-input agricultural systems.

CALCIUM, pH, AND SOME RELATED FACTORS

Freire (1976) reviewed the effects of soil acidity and pH-related factors on soybean nodulation and N_2 fixation. More recent reviews have also appeared (Munns, 1976; Munns, et al., 1977) in which important new findings on tropical and temperate legumes are

reported, findings that permit a better understanding of their behavior in acid soils.

Growth and activity of Rhizobium as well as effective nodulation requires an adequate supply of Ca. However, it is difficult to
separate the nutritional effects of added Ca and the neutralizing
influence of $CaCO_3$. The role of Ca in symbiotic N_2 fixation is well
documented by de Mooy et al., and by Foy (Freire, 1976).

There is some correlation between soil acidity and the availability to plants of such macro- and micronutrients as Ca, P, Mn,
Al, B, Mo, and Cu. Lime applications correct Ca deficiency and the
toxicities of Al and Mn, and it also reduces or increases the availability of many other elements and nutrients in soil. These effects
occur by biological mechanisms, chemical mechanisms, or both. It is
understandable that the close relationship between Rhizobium, host,
and environment (including soil factors) is affected by soil acidity,
and any factor that changes soil acidity will also affect the relationship.

Soil acidity influences N_2 fixation by direct and indirect
effects on the bacteria and on the host. Rhizobium species and
strains vary in their tolerance to soil pH. According to recent
research, the dogma that tropical legumes and their rhizobia are acid
tolerant and temperate legumes and their rhizobia are not tolerant
to acid conditions is no longer valid. Norris proposed that tropical legumes (including soybeans) are ancestors of the family Leguminosae and evolved together with their bacterial symbionts in the acid
soils of low fertility which predominated in the Cretaceous period
(Freire, 1976). Thus, tropical legumes were presumed to have developed and maintained a high capacity for extracting nutrients,
including Ca, from the soil, and the associated bacteria that evolved
were those that produced alkaline products, which tend to neutralize
the acid environment; these bacteria grow slowly in culture media.
In fact, the temperate legumes that have evolved are more adapted to
highly fertile, nonacid soils, and their associated rhizobia produce
acid and grow rapidly. From a practical standpoint and when one
considers the better behavior, growth, nodulation, and establishment of most tropical legumes in comparison with most temperate
legumes in acid soils of low fertility, Norris' hypothesis seemed
valid. However, this hypothesis has been subjected to severe cri-
ticism. Munns (1976) and Munns et al. (1977), based on recent papers
and their own work, concluded that the behavior of one legume in
acid soil cannot be deduced from experience with another. In other
words, among the tropical and temperate legumes, one can find both
tolerant and intolerant species and hence different responses to
lime. Even within one species, variations can occur, depending on
its adaptation to environmental conditions. Munns (1976) presented
a list of 29 legume species categorized according to their tolerance
to soil acidity (Table 2). In one extreme are Medicago sativa and

TABLE 2. Relative Acid Tolerance of Legume Species[a] (a first approximation)

Species	Nodulation	Growth[b]	Species	Nodulation	Growth[b]
Acacia mellifera	2	?	Medicago sativa	1	1
Arachis hypogea	4	4	Medicago truncatula	1	1
Centrosema spp.	4	4	Medicago scutellata	1	1
Coronilla varia	0	0	Phaseolus vulgaris		
Desmodium canum	3	3	Black-seeded	3	3
Desmodium intortum	3	3	Other	2	2
Desmodium uncinatum	4	3	Pisum sativum	2	2
Glycine max	3	3	Pueraria spp.	?	5?
Glycine wightii	2	2	Stylosanthes humilis	5	4
Indigofera spicata	4	4	Trifolium africanum	4	3
Leucaena latisiligua	4?	0	Trifolium pratense	1	2
Lotononis bainesii	5	4	Trifolium repens	1	2
Lotus corniculatus	2	2	Trifolium subterraneum	2	3
Macroptilium lathyro	5	5	Trifolium ruepellianum	3	3
Macrotyloma axillare	4	2	Trifolium semipilosum	3	3
			Vigna unguiculata	5	4

[a] Rating scale from 0 (most intolerant of soil acidity) to 5 (most tolerant).
[b] Growth when dependent on nitrogen fixation.

other <u>Medicago</u> species; at the other extreme are <u>Lotononis bainesii</u> and <u>Stylosanthes humilis</u>. The lack of response to liming by <u>L. bainesii</u> has also been observed by Scholles et al. (1981) in an acid soil of southern Brazil. Munns (1977) observed that the use of legumes and rhizobia that are adapted to soil acidity reduces the need for expensive inputs of lime and that some important tropical types that evolved or developed in areas of neutral, alkaline, or calcic soil may contain types that are not adapted to acid soils. On the other hand, the tolerance to soil acidity that evolved or developed might not have been associated with a vigorous N_2-fixation capacity. Good examples of successful adaptation and selection are <u>Medicago</u> species for nonacid environments and <u>Trifolium subterraneum</u> for acid soils. Kornelius and Saibro (1973) reported better nodulation, hay yield, total N, and seed production of <u>T. subterraneum</u> at pH 5.5 (3 tons of lime per ha), and they noted that liming favored survival of the inoculated rhizobia. Some <u>Rhizobium</u> strains are more apt to survive at pH 5.0. Stammel (1968) found that in Al-rich soils, liming to pH 6.0 to 7.0 had a major effect on the yield of red clover (<u>Trifolium pratense</u>) in the state of Rio Grande do Sul, Brazil. Kolling (1974) obtained the maximum yield of siratro at pH 5.5 (using three-fourths of the amount of lime recommended to reach pH 6.5) and of <u>Desmodium</u> at pH 5.9 (using the recommended rate to reach pH 6.5).

The sensitivity of alfalfa to acidity is a major factor limiting the widespread introduction of this forage into the area of oxisols that predominate in southern Brazil. Silva (1973) obtained higher alfalfa yields with 14 tons of lime per ha in an alfisol of pH 5.2 and found a significant correlation between nodule weight, yield of hay, and total N per ha (Table 3).

Pons et al. (1974) obtained a response to liminh with up to 40 tons per ha two years after the lime application. In practical terms, soil analysis laboratories recommend lime for alfalfa planted in the state of Rio Grande do Sul at twice the usual rate to reach pH 6.0. This rate is 8000 to 10,000 tons of lime per ha, and despite the expenses involved, alfalfa is being more widely grown in the oxisols and farmers are obtaining an economic return.

<u>Phaseolus vulgaris</u> is another example of adaptation to nonacid or slightly acid soils of high fertility. This adaptation was natural, but it also occurred as a consequence of the selection and breeding that is usually conducted in soils of high fertility. The symbiotic capacity was lost or not developed because there was no attention to its development. In Brazil and other countries in Latin America, the traditional cultivation of beans is a small farmer's enterprise and takes place in the rich valleys, newly opened forest areas, or on steep hills of young brunizem and lithosols that are fertile and usually rich in organic matter. In a survey of traditional bean-growing areas of Rio Grande do Sul, Pons

TABLE 3. Effect of Lime*, P, and K on Yield and Weight of Nodules in the "Alto das Canas" Soil, Brazil (from Silva, 1973)[a]

K treatment	Factor measured	Ca 0			Ca 1			Ca 2		
		P0	P1	P2	Po	P1	P2	P0	P1	P2
K0	Yield[b]	0.5	1.3	1.4	2.9	6.7	7.4	5.5	8.0	9.1
	Nodulation[c]	8	7.0	0	23	127	199	13	154	213
K1	Yield	0.7	1.9	1.8	2.8	7.9	9.6	7.3	9.7	10.1
	Nodulation	0	0	23	12	133	276	47	144	163

[a]Treatments: P1, P2, K1, Ca 1, and Ca 2 denote 100 and 200 kg P_2O_5 per ha, 200 kg K_2O per ha and 7 and 14 tons lime per ha, respectively. The soil pH was 5.2, 5.9, and 6.4 for treatments Ca 0, Ca 1, and Ca 2, respectively.
[b]Tons dry matter per ha.
[c]mg per plant.

et al. (1976) observed that good (natural) nodulation occurred only
in the soils that are naturally highly fertile. Graham and Halliday
(1976) reported that soil temperature is a major limiting factor for
beans in tropical and substropical areas, the limiting pH for growth
of R. phaseoli in liquid medium was 4.0 to 4.4, and according to the
work of Spain et al., black-seeded cultivars show some tolerance to
soil acidity. Voss (1981) studied 60 R. phaseoli isolates from bean
soils of the state of Rio Grande do Sul and reported that pH and related factors had a marked effect on the comparative ability of 4
strains. P. vulgaris is adapted to soils with pH values of 6.0 or
higher and which are rich in P, K, Ca, and Mg and have no exchangeable Al. However, attempts to obtain good nodulation of beans in
clay oxisols by correcting the acidity and improving the nutrient
level have not been successful and the responses have been nil or
erratic. It seems that the factors limiting the formation of nodules
on beans are still not well understood. There is no doubt that a pH
close to neutrality is important for R. phaseoli. However, one can
correct the pH of the soil (mainly clays) and supply water and all
macro- and micronutrients, but still there is no guarantee of good
nodulation in pots or in the field, as occurs with, for example,
soybeans.

Halliday and Abelli (1981), in a study of 100 isolates of
Rhizobium from Arachis hypogaea, Desmodium intortum, Leucaena leucocephala, Medicago sativa, Phaseolus vulgaris, and Stylosanthes
guianesis, reported that when primary isolates were made onto a medium at pH 4.5, a few strains grew better at this pH and did not
grow at pH 7.0. There was an indication that rhizobia are more
tolerant of alkalinity than acidity and that the relative distribution of pH tolerance depends on the legume species from which the
Rhizobium strains were isolated and the pH of the soil from which
the nodules were collected. They pointed out that most attempts to
screen rhizobia for acid tolerance involved strains from existing
culture collections and so the primary isolations were made in the
traditional media with neutral pH values. For this reason, the
strains really tolerant to acid conditions were probably not isolated or the tolerance had been lost during prolonged cultivation
in media of neutral pH value. Their results may explain in part the
existing controversy on the pH requirements for growth of rhizobia
in agar or liquid media. This work confirmed the results of Date
and Halliday (1979) with isolates from Stylosanthes spp.

Another explanation for the controversey, according to Munns
(1976), is the difficulty controlling pH and possible pH interactions with other factors. These difficulties have now been overcome
by the acid medium developed by Date and Halliday (1979) that will
make it possible to test the effects of characteristic conditions
of acid soils.

TABLE 4. Effect of Lime and Strain of R. trifolii on Plant Yield and Weight of Nodules of Trifolium subterraneum in Two Periods of Growth (from Kornelius et al., 1972)[a]

R. trifolii	Period of growth	0 t/ha[b]		1.2 t/ha[b]		2.4 t/ha[b]	
		DM[c]	Nod[d]	DM[c]	Nod[d]	DM[c]	Nod[d]
None	1970	0.65	0	0.71	0	1.62	0
	1971	0.66		0.54	0	0.54	0
Strain 222	1970	2.38	114	2.96	99	3.99	104
	1971	0.47	0	4.58	141	4.78	85
Strain 235	1970	2.97	114	3.76	120	4.12	146
	1971	0.91	20	2.76	84	4.00	91
Strain 239	1970	2.22	92	2.83	98	3.40	99
	1971	0.46	0	3.77	87	5.11	115
Strain 265	1970	171	103	2.04	109	2.84	123
	1971	0.48	0	4.01	198	4.74	159

[a] Ineffective small nodules from parasitic native strain were not considered.
[b] Lime addition.
[c] Dry matter in g per pot.
[d] Nodule dry weight in mg per pot.

It seems that the data are more certain with regard to R. meliloti, which is less tolerant to low pH. Different serogroups of R. japonicum, however, vary greatly. Munns (1977) observed that rhizobial colonization and survival in conventional media or in soil differs from its behavior in the host rhizosphere. The number of clover Rhizobium in an acid soil increased with the addition of CaCO₃. It is well accepted that the Ca requirement for growth of rhizobia is lower than for nodule formation and for plant growth. However, plant uptake can reduce Ca availability to a point at which rhizobial growth would be affected. Changes in the environment will affect the population of all soil microorganisms as well as rhizobia. However, a few data have been presented showing the influence of the environment on rhizobial populations.

In an acid soil rich in a native Rhizobium that nodulates efficiently a native clover (Trifolium polymorphum) but is parasitic for Trifolium subterraneum, Kornelius, Freire, and Barreto (1972) were able to introduce efficient strains for subterranean clover by liming the soil to pH 6.5. The lime vs strain interaction was still highly significant in the second period of growth, and this was an indication of the importance of lime to survival of the introduced rhizobia (Table 4).

The effect of soil acidity on nodulation and N₂ fixation can be a consequence of a direct inhibition of Rhizobium survival, colonization, or competitiveness for nodule sites. Such effects would explain the findings of several authors, as cited by Munns (1976) and Freire (1976), that inhibition of nodulation has been overcome partly in some experiments by increasing the number of cells in the inoculum. An alternative explanation, according to Munns, is that acidity also causes an inhibition of rhizobial growth and thus results in a smaller population. In a Brazilian acid oxisol exhibiting Al and Mn toxicity, an improvement of nodulation by increasing the size of the inoculum was also observed (Freire et al., 1974). However, in spite of the clear indication of this stress, it is common to use poor inoculants in acid soils (and in other stress conditions) without regard to the rhizobial numbers in the inoculant. In the "cerrado" region of central Brazil, the areas are usually brought into cultivation by planting rice for one to two years, and then soybean is planted. It is presently recommended that the problem of poor nodulation of soybeans in the first year can be overcome by inoculating the rice seeds at the time of their first plantings; when soybean is planted, it encounters an established population of rhizobia in the soil. This procedure is being used by farmers.

In many of the acid soils of the tropics and subtropics and even in some temperate soils, toxic levels of exchangeable Al, Mn, or both are present. Many workers have reviewed the extensive literature and/or presented original results on the effects of Al

and Mn toxicities on the Rhizobium-legumes symbiosis (Freire, 1976).
The effect of Al toxicity appears to be indirect; e.g., on root
growth, nodule initiation, or on the physiology of the host plant.
Aluminum interferes with the uptake and use of Ca, and plant tolerance to Al seems to be closely associated with the ability to absorb
and transport Ca in the presence of excess Al. The inhibitory effect
of Al on Ca absorption is well confirmed by findings related to the
Al:Ca:Mn relationshp. Tolerance to Mn toxicity is related to high
Ca:Mn ratios in plant tissues rather than to the content of Mn alone.
Munns (1976) cites critical concentrations of Mn in tissues of many
tropical and temperate legumes. The figures presented ranged from
200 ppm for soybeans to 1600 ppm for Centrosema pubescens.

There is also evidence that Al tolerance is related to the ability of the plant to absorb and utilize P in the presence of Al.
Several studies on the effect of lime and P on nodulation, N_2 fixation, and yield of soybeans conducted in acid soils in southern
Brazil were reviewed by Freire (1976). The neutralization of high
acidity and such linked factors as Al toxicity and the corection of
nutritional deficiences are essential to the attainment of high N_2
fixation and high productivity. In some soils, P fertilization
per se alleviates the inhibition of nodulation, but where toxic
levels of Al and/or Mn are present, liming (high lime levels being
needed in some soils) is essential to promote adequate nodulation
for high N_2 fixation and high yield.

The direct effect of Al and Mn toxicities on rhizobia was
studied by Keyser and Munns (1979). They reported that in acid soils,
Al toxicity and acidity itself were probably more important factors
limiting rhizobial growth than Mn toxicity and Ca deficiency.

There has been little study of the tolerance of legumes to Al
and Mn toxicity. The research done so far has usually dealt with
one of the elements but not with both. It seems that insensitive
or less sensitive species to Al or Mn toxicity are more common among
tropical legumes than among temperate legumes. Desmodium uncinatum,
Lotononis bainesii, Macroptilium lathyroides, and Trifolium rueppel-
lianum are cited by Munns (1976) as very insensitive to Al toxicity.
Trifolium subterraneum, T. semipilosum, T. repens, and Glycine wightii
are of intermediate sensitivity, and Medicago sativa and M. scutellata
are sensitive. Kamprath and Foy (cited by Freire, 1976) claim that
temperate legumes are more sensitive to Mn toxicity than are tropical
legumes. However, it seems that this generalization is not supported
by a sufficient number of studies with tropical species.

According to several recent papers, an important finding of the
past decade is that soybeans do not necessarily require high pH for
optimum growth (Freire, 1976). It was earlier believed that soybeans had an optimum pH between 6.5 and 7.0. The pH per se is not
a major limiting factor for soybeans. Good soybean nodulation and

high yields have been observed even at pH 4.5 to 5.0. However, if soil acidity is associated with Al and/or Mn toxicity, an inhibition of nodulation, N₂ fixation, and other facets of plant function occurs, and the yield is reduced.

Mengel and Kamprath (1978) reported that the critical pH for most of the soils they stidied (in North Carolina, United States) was in the range of pH 4.6 to 4.8 and that nodule number and weight and the N content of plant tissue increased significantly as the soil pH increased.

There is a large variation among soybean cultivars in respect to the tolerance to soil acidity and to Al and Mn toxicities. There is evidence that cultivars that absorb more Ca are more tolerant to Al and Mn toxicities. This observation is in accord with the reported relationship Ca:Al:P:Mn mentioned in this paper. Coutinho et al. (1975) in a study involving six soybean cultivars and two lime levels reported that the tolerant cultivars absorbed more Ca and less Mn than the nontolerant cultivars.

The effects of soil acidity on the availability of macronutrients and micronutrients and on exchangeable Al and Mn are important agronomically, and they deserve considerable attention in the less developed countries. Based on the literature dealing with problems in Australian soils, we believe that there has been widespread promotion and application of some "panaceas" in some areas of South America and in other regions. First, superphosphate was promoted as a means to solve all soil fertility problems. Then came the promotion of Mo as the way to have high N₂ fixation everywhere and the promotion of seed pelleting to overcome soil acidity problems. In some areas, because of the lack of adequate local research, these "panaceas" did not help to increase yields but instead brought frustations. Seed pelleting and Mo are certainly beneficial in acid soils but not in those with toxic levels of Al and/or Mn. Research for the identification, quantification, and correction of limiting factors in soil together with economical analyses are needed everywhere. Once the responses to lime and fertilizers are established for each type of soil, economical analyses will allow for the formulation of recommended practices at any time or for various economic conditions.

POTASSIUM

The effect of K on the Rhizobium-legume symbiosis is indirect and acts through the physiology of the plant because there is no direct effect on the bacteria or on nodule formation.

Andrew (1976) reviewed the influence of K on legume nutrition and observed how important is this element to the long term maintenance

of legume-based pastures, particularly in areas with medium to high
rainfall.

The level of available K is commonly high in many soils of the
tropics and subtropics, even in soils of low fertility and soils that
are underutilized because they are in low productivity agriculture
or in natural pastures. Soil analyses usually show high levels of
K availability, and there is no response to K in short-term fertility trials. However, when these trials are conducted for more than
two years, K deficiency begins to appear.

An effect of K on nodulation and yield of soybeans was observed
by Kolling et al. (1974), and it was evident in the third year of
growth of three soybean cultivars. The results presented in Table
1 refer to the fourth year of planting.

Because of the high yields that have been obtained in some
areas in southern Brazil by liming and fertilizers and because of
the large amounts of K removed from the soil, K deficiency is now
appearing in many fields because of insufficient fertilization with
K.

Potassium may also be detrimental to nodulation and N_2 fixation. Borkert (cited by Freire, 1976) observed that the application of KCl without lime increased the availability of Mn in soil
as well as the Mn concentration in soybean tissue. Nodulation was
reduced in plants grown in pots but not in the field in Brazilian
acid soil rich in Al and Mn. The same effects were observed in a
study of alfalfa grown in six soils: KCl at lower lime rates reduced the pH and increased the level of exchangeable Mn in the soil
and the Mn concentration of the plant tissue (Kornelius, 1972). Reduction of nodulation in the field was also observed with soybeans.
The detrimental effects have been attributed to the Cl ion. This
indicates the danger of applying KCl without lime in acid soils with
high exchangeable MN (Freire, 1976).

SULFUR AND MICRONUTRIENTS

Several authors have reviewed the literature on the role of
sulfur and micronutrients on the Rhizobium-legume symbiosis (Andrew,
1976; Freire, 1976; Franco, 1976).

The importance of sulfur results from its role as a protein
component. Sulfur deficiency has been observed in central and
southern Brazil. In a pot experiment reported by Borges et al.
(1974) with soybeans in sandy and clay soils, there were positive
responses to applications of S. The S increased the nodule weight
more in the absence of lime than with lime. Stammel et al. (1976)
also observed the influence of sulfur with and without liming on

alfalfa production. Andrew (1976) refers to the existence of S-responsive areas in northern Australia. He states that there is a close relationship between N and S concentratins in tropical legumes. In the state of Sao Paulo, Brazil, Freitas (1969) obtained significant increases in alfalfa and Stylosanthes gracilis yields by applications of S.

Zinc is another micronutrient that may be deficient in soils of high pH and in soils heavily weathered, as in the "cerrado" soils of South America. Andrew (1976) reviewed the extensive literature on Zn; nevertheless, field experimentation on the effects of Zn on nodulation and N_2 fixation is meager. Anghinoni et al. (1976) in a field experiment and Borges (1974) in a pot experiment found no responses of soybeans to applications of Zn. However, micronutrient deficiencies have to be sought in long-term field experimentation with high nutrient uptakes.

MOLYBDENUM AND BORON

Molybdenum is important because it is part of the nitrogenase molecule. Its role in N_2 fixation has been well documented. However, it can also be detrimental when in excess amounts. Positive responses to Mo application in the field have been obtained by seed treatment, soil treatment, or foliar application in soils with low pH values and nontoxic levels of exchangeable Al and/or Mn. In soils that have been limed or in soils with high pH values, the availability of Mo is usually adequate for N_2 fixation, and there is no response to Mo applications. Some experiments, however, have shown positive responses in limed soils. These results may result from the existence of toxic levels of Al and/or Mn, the use of small amounts of lime or the presence of very little Mo in the soil. It has been also observed that Phaseolus vulgaris has a very high requirement for Mo (Franco, 1976). Molybdenum compounds applied to the seed together with the inoculant can be detrimental to bacterial survival.

Boron is required for the normal development of roots and nodules. It is involved in protein metabolism, but it is not essential for rhizobia. Positive responses have been obtained in several experiments. In the state of Rio Grande do Sul, Brazil, 20 kg/ha/year of boron is recommended by soil analysis laboratories when the soil is limed for alfalfa planting (Stammel et al., 1976). In the traditional alfalfa areas in fertile soils, B deficiency has not been observed. However, in a field experiment in an oxisol of southern Brazil, B application significantly decreased yield but did not affect the dry weight of nodules (Anghinoni et al., 1976). Boron toxicity had been observed previously in this and other oxisols in pot experiments in which micronutrients were applied using usual micronutrient formulations.

TABLE 5.   Dry Matter of Plant Tops, Roots, and Nodules of Phaseolus
           vulgaris Planted in Pots with an Inside Cylinder of Sand
           or Soil and an Outside of Soil or Sand (from Kornelius
           et al., 1972)

| Treatment[a] | | Dry matter yield (g/pot) | | |
Outside	Inside	Plant tops	Roots	Nodules
13.0SL	0.0SD	0.50	0.46	0
8.5SL	4.5SD	0.70	0.66	0
9.5SL	3.5SD	0.78	0.76	0
10.5SL	2.5SD	0.96	1.07	0
11.5SL	1.5SD	1.53	1.05	0
0.0	13	2.45	1.40	0.34
8.5SD	4.5SL	2.99	1.38	0.22
9.5SD	3.5SL	3.90	1.84	0.17
10.5SD	2.5SL	2.09	1.34	0
11.5SD	1.5SL	1.34	1.03	0

[a]SL, soil. SD, sand. The diameter of the pot cylinders was 13 cm.

SOIL MOISTURE AND AERATION

     The impact of soil moisture and aeration on nodule formation
under field conditions has not received appropriate attention. Most
of the work reported so far has been on the effects of moisture and
drought conditions on nodulation and the process of N_2 fixation in
artificial conditions (Sprent, 1976; Gibson, 1976).

     Diatloff (1967) observed that during wet periods in a black
earth soil, aeration was the limiting factor for nodulation of cowpeas, soybeans, and native legumes. Coepfert and Freire (1973) obtained significant increases in nodulation and dry matter of Phaseolus
vulgaris in a soil sieved to obtain particles of 0.8 to 2.0 mm. An
oxisol high in Al and Mn was used, but it was limed to pH 6.5 and
fertilized. Plants in the treatments with no improved aeration had
no nodules. The absence of nodules or poor nodulation is a common
observation in pot experiments with clay soils. On the contrary,
good nodulation always occurs in pots of sand or in Leonard jars with
sand. Kornelius and Freire (1974) established a pot experiment in
which each pot (diameter of 13 cm) had a cylinder of soil or sand in
the center and sand or soil on the outside. Water was added to maintain a constant weight and to maintain 80% of the field capacity of
the soil and the sand. It was expected that nodules would be formed
only when the inside cylinder was sand and not when it was soil.

TABLE 6.   Effect of Moisture on Weight of Nodules, Plant Weight, and Total N of 45-day-old Plants (adapted from Saito et al., 1979)

Water tension (atm)	Nodules (mg dry wt/pot)	Plant (g dry wt/pot)	Total N (mg/pot)
15	33.8	3.5	88.5
1.0	207.0	5.8	159.7
0.06	298.5	7.0	201.3

However, nodulation occurred when the inside cylinder was soil at the two higher diameters and not when the inside cylinder was soil. The soil absorbed water from the surrounding sand and its aeration was also improved by the sand. As expected, nodulation was good in the control pots containing sand. Dry matter of the plant tops and roots was not related to effects on nodulation (Table 5).

Saito et al. (1979) established an experiment with pots of soil with controlled water tensions to study the effect of moisture and ^{15}N fertilizer on nodulation and N_2 fixation by Phaseolus vulgaris. The fertilizer was $(NH_4)_2SO_4$. Plants cultivated in wet soils (water potential of 0.06 atm) had dry matter, total N, and P levels of shoots two times greater than those from dry soils (water potential of up to 15 atm). Weight and nitrogenase activity of the nodules from plants in the wet soil were 5 to 10 times higher than nodules from the dry soil. N fertilizer inhibited nodulation, but this effect was less in wet soil.

Bonetti and Saito (unpublished data) reported that the efficiency of P uptake increased with soil water content, and they noted a positive interaction between P in the soil, water content, and nodulation.

It seems that moisture is a critical factor limiting nodulation of Phaseolus vulgaris. It has been commonly observed in the field that soybeans, for example, nodulates well in dry soil conditions in which beans do not form nodules. The critical moisture range for nodulation of beans seems to be narrower than that for soybeans and other legumes. This fact would explain why the results of bean inoculation trials in the field are erratic and also why good native nodulation is commonly found in sandy black soils rich in organic matter. These soils would maintain adequate moisture and aeration for bean nodulation for a longer time than would clay or sandy soils. The use of soil conditioners, straw, or other organic residues may

be a way to overcome moisture stress on the nodulation of beans.
However, Selbach et al. (1980) reported a detrimental effect of corn
residues on nodulation of Phaseolus vulgaris. Positive responses
were obtained with chicken bedding and cattle manure. Soils with
these treatments had higher levels of ammonium and nitrate than the
control soil and lower levels were present with corn straw. Rice-
hull mulch increased nodulation and N_2 fixation by beans (Waters
et al., 1980). Rice-hull mulch increased the yield of three soybean cultivars from 25 to 126% (Freire, 1965). These effects result from the increase in the water-holding capacity of the soil
and the evolution of CO_2 from decomposition of the organic matter.
The work of Shivashankar and Vlassak (1978) confirmed these results.
They obtained a marked increase in yield and nitrogenase activity of
soybeans by the use of straw (2 to 4 t/ha) with or without CO_2 enrichment. Nitrogenase activity was increased by 141 to 197% by
straw addition to the soil, and CO_2 enrichment increased the activity by 24%. Straw alone (4 t/ha) or in conjunction with CO_2 increased the activity four-fold. The authors concluded that straw
could be a partial substitute for CO_2 enrichment for improving N_2
fixation, growth, and yield of crops. It seems that different results have been obtained in beans and in soybeans by the use of organic residues. This should have been expected because of the differences in plant-tissue components and in the periods of decomposition. Unfortunately, additional data are needed to allow for an
adequate interpretation of these results.

COMBINED NITROGEN

     Freire (1976) and Franco (1976) reviewed the literature on the
effect of mineral nitrogen in soil on the Rhizobium-legume symbiosis.
If nitrate or ammonium is in excess, detrimental effects occur because of an inhibition of nodule formation and nitrogenase activity.
However, legume species and cultivars and also Rhizobium strains
vary widely in their tolerance to combined nitrogen in soil. The
search for bacteria and plant types adapted to high levels of mineral nitrogen in soil is a feasible way to attain higher yields,
for example, of soybeans.

     Mineral nitrogen in small amounts is beneficial to nodulation.
Franco (1976) cited several authors who obtained increased nodulation at the flowering stage of the plant. The same author also observed that the initial nodulation and onset of N_2 fixation by some
legumes in sand and N-free nutrient solution are stimulated by
a small amount of N. With Phaseolus vulgaris, if this small amount
of N is not added, nodule formation, N_2 fixation, or both are delayed considerably, and N_2 fixation is poor up to the flowering stage
and to the end of plant growth.

The effect of supplmental N in raising the yield of soybeans has been investigated by a large number of researchers. However, according to Franco (1976), positive responses to nitrogen applications have not usually been accompanied by data on grain yields.

According to Freire (1976), the following important conclusions can be drawn. (a) There is an inverse relationship between mineral N and symbiotic nitrogen, a relationship that may be explained by a competition for photosynthates. (b) Positive responses to applied N may be obtained in stress conditions or during poor-growing seasons, at which time nodulation, N_2 fixation, or both would be inhibited. The prevailing environmental conditions are responsible for the variable results and the controversy. The N requirement of soybeans under field conditions is more or less well satisfied by soil N and N_2 fixation. However, detrimental moisture conditions, for example, would limit the mineralization of organic matter and nitrification, and so a positive response to N fertilizer may occur mainly at two critical periods: during rapid vegetative growth and during pod filling. Weber found positive responses in 7 of 21 experiments, and Welch found responses in 3 of 130 instances using different rates, times, and methods of application and with organic or inorganic N sources and in different locations (Freire, 1976). The three positive responses occurred at uneconomically high rates of fertilizer use. (c) Plants supplemented with mineral N at planting time receive a better start and are taller and greener because N_2 fixation starts 10 to 14 days after germination. However, inoculated plants with no N recover rapidly so the grain yield is similar. (d) Prolonged adverse environmental conditions may increase the influence of added N. (e) Poor nodulation or nodulation by combinations of Rhizobium strain x soybean cultivar that have low efficiencies may also account for the response to N fertilizer. (f) Nitrate fertilizer is more detrimental to nodulation than ammonium. Urea seems to be less detrimental than nitrate or ammonium. (g) Nitrate uptake and metabolism are of primary importance during vegetative growth and flowering, whereas N_2 fixation becomes more important during pod filling. Franco (1976) also observed that the contribution of N from N_2 fixation is higher and from nitrate metabolism is lower after flowering than during vegetative growth. (h) The mechanisms of the reduction of the symbiotic process in the presence of combined nitrogen is still not well understood. (i) Deep placement of N fertilizer is less detrimental to nodule development than uniform distribution. (j) In acid soils rich in exchangeable Al and Mn, positive responses to N fertilizer may also occur. Liming increases the availability of mineral N and neutralizes the toxicity to nodulation and N_2 fixation.

A large number of experiments on the use of supplemental N on soybeans have been run in Brazil, and only a few positive responses were recorded and these were some time ago. However, the pressure from the fertilizer companies is high in spite of the recommendations by research and extension organizations against the application of N fertilizers to soybeans. Camp and Sfredo (1981) estimated that 85,000 tons of N is being applied in soybean fields every year through the "traditional" formulas for fertilizers, which contain 10 to 15 kg of N as "starter" for early growth. All the experiments that have been performed show that this "starter" is useless and expensive.

SUMMARY

The close relationship between legume, Rhizobium, and the environment governs N_2 fixation by the symbiotic association. If the requirements of the host and of the bacteria are not satisfied, the extent of fixation and productivity of the plant are reduced. The more important and common limiting factors in soil for N_2 fixation in developing countries of the tropics are P deficiency, soil acidity, and toxic levels of exchangeable Al and Mn. The overcoming of these stresses is essential for high N_2 fixation and high yields. The search for and use of legumes and of their specific bacteria that tolerate these stress conditions is important but research information is also needed on practices for completely correcting the limiting factors and for the attainment of high yields.

The effects of P deficiency have been repeatedly reported for temperate and tropical forage legumes, soybeans, and common beans. Soil acidity, exchangeable Al and Mn, and P and Ca deficiencies represent a complex relationship that can seriously limit nodulation and N_2 fixation.

There is no relationship between the origin of legumes and their tolerance to soil acidity and the related factors, such as toxic levels of Al and Mn, although tolerance seems more common among the tropical species. Evidence has been presented that the tolerance of Rhizobium is linked to adaptation of its host.

Potassium, S, and micronutrients in soil have minor roles as limiting factors, at least in areas of low-productivity agriculture. At least for Phaseolus vulgaris, soil moisture seems to be an important factor limiting nodulation and N_2 fixation.

The inhibitory effect of combined N on nodulation and N_2 fixation is well established. However, there is a large and widespread application of fertilizer N that, at least for soybeans, is useless and wasteful, and such practices reduce the economic return to farmers.

REFERENCES

C. S. Andrew, in: Exploiting the Legume-Rhizobium Symbiosis in
    Tropical Agriculture (A. S. Whitney and E. Bose, eds.), Univ.
    of Hawaii, Honolulu, pp. 253-274 (1977).
I. Anghinoni, I. Fiorese, and A. P. Moraes, Agron. Sulriogr., 12:
    189-199 (1976).
M. T. Beaty et al., J. Agron. Educ., 1:37-40 (1972).
R. Bonetti and S. M. T. Saito (Unpublished data).
A. C. Borges et al., in: VII Reunion Latinoamericana Sobre Rhizo-
    bium, Resistencia, Argentina, pp. 40-47 (1974).
R. J. Camp and G. J. Sfredo, EMBRAPA Centro Nacional de Pesquisa da
    Soja, Comunicado Tecnico, No. 8 (1981).
C. Coutinho, J. R. J. Freire, and C. Vidor, in: Proc. IVa Reuniao
    Latinoamericana sobre Rhizobium, Rio de Janeiro, p. 169 (1975).
R. Date and J. Halliday, Nature (London), 277:62-64 (1974).
A. Diatloff, Queensland J. Agric. Anim. Sci., 24:315-321 (1967).
E. L. Fox and B. T. Kang, in: Exploiting the Legume-Rhizobium
    Symbiosis in Tropical Agriculture (J. M. Vincent, A. S. Whitney,
    and E. Bose, eds.), Univ. of Hawaii, Honolulu, pp. 183-210
    (1977).
A. A. Franco, in: Exploiting the Legume-Rhizobium Symbiosis in
    Tropical Agriculture (J. M. Vincent, A. S. Whitney, and E. Bose,
    eds.), Univ. of Hawaii, Honolulu, pp. 273-274 (1977).
J. R. J. Freire, in: Exploiting the Legume-Rhizobium Symbiosis in
    Tropical Agriculture (J. M. Vincent, A. S. Whitney, and E. Bose,
    eds.), Univ. of Hawaii, Honolulu, pp. 335-379 (1977).
J. R. J. Freire, in: Proc. II Reuniao Latinoamericana Sobre Rhizo-
    bium, IDIA, Suppl, Buenos Aires, pp. 33-38 (1965).
J. R. J. Freire, et al., in: Proc. VII Reunion Latinoamericana Sobre
    Rhizobium, Resistencia, Argentina, pp. 124-129 (1974).
L. M. M. Freitas, Unpublished data (1969).
A. H. Gibson, in: Symbiotic Nitrogen Fixation in Plants, (P. S.
    Nutman, ed.), Cambridge Univ. Press, London, pp. 385-402 (1976).
C. F. Goepfert, Agron. Sulriogr., 7:5-9 (1971).
C. F. Goepfert and J. R. J. Freire, Agron. Sulfriogr., 8:187-193
    (1972).
C. F. Goepfert and J. R. J. Freire, Agron. Sulfriogr., 9:143-149
    (1973).
P. H. Graham and J. Halliday, in: Exploting the Legume-Rhizobium
    Symbiosis in Tropical Agriculture (J. M. Vincent, A. S. Whitney,
    and E. Bose, eds.), Univ. of Hawaii, Honolulu, pp. 313-334
    (1977).
J. Halliday and C. Abelli, in: Biological Nitrogen Fixation Tech-
    nology for Tropical Agriculture, CIAT, Cali, Colombia (1981).
H. H. Keyser and D. N. Munns, Soil Sci. Soc. Am. J., 43:500-503
    (1979).
J. Kolling, M.Sc. Thesis, Faculdade de Agronomia, UFRGS, Brazil
    (1974).

J. Kolling et al., in: Proc. VII Reunion Latinoamericana Sobre
    Rhizobium, Resistencia, Argentina, pp. 342-353 (1974).
E. Kornelius, M.Sc. Thesis, Faculdade de Agronomia, UFRGS, Brazil
    (1972).
E. Kornelius and J. R. J. Freire, Agron. Sulriogr., 10:274-260
    (1974).
E. Kornelius, J. R. J. Freire, and I. L. Barreto, Agron. Sulriogr.,
    8:95-109 (1972).
E. Kornelius and J. C. Saibro, Agron. Sulriogr., 9:57-70 (1973).
D. B. Mengel and E. J. Kamprath, Agron. J., 70:959-963 (1978).
B. Mosse, in: Exploiting the Legume-Rhizobium Symbiosis in Tropi-
    cal Agriculture (J. M. Vincent, A. S. Whitney, and E. Bose,
    eds.), Univ. of Hawaii, Honolulu, pp. 275-292 (1977).
D. N. Munns, Exploiting the Legume-Rhizobium Symbiosis in Tropical
    Agriculture (J. M. Vincent, A. S. Whitney, and E. Bose, eds.),
    Univ. of Hawaii, Honolulu, pp. 211-236 (1977).
D. N. Munns, R. L. Fox, and B. L. Koch, Influence of lime on nitro-
    gen fixation by tropical and temperate legumes, Plant Soil,
    46:591-601 (1977).
A. L. Pons et al., Agron. Sulriogr., 10:211-226 (1974).
A. L. Pons et al., Agron. Sulriogr., 12:129-132 (1976).
S. M. T. Saito, et al., XVII Congr. Bras. Ci. Solo (1979).
D. Scholles et al., Agron. Sulriogr., 14:303-310 (1978).
D. Scholles et al., Agron. Sulriogr., 14:135-142 (1978).
D. Scholles, J. Kolling, and J. R. Freire, R. Bras. Ci. Solo, 5:
    97-102 (1981).
P. A. Selbach, D. Scholles, and J. Kolling, Unpublished data (1980).
K. Shivashankar and K. Vlassak, Plant Soil, 49:259-266 (1978).
V. P. S. Silva, M.Sc. Thesis, Faculdade de Agronomia, UFRGS, Brazil
    (1973).
J. I. Sprent, in: Symbiotic Nitrogen Fixation in Plants (P. S.
    Nutman, ed.), Cambridge Univ. Press, London, pp. 405-420
    (1976).
J. G. Stammel, M.Sc. Thesis, Faculdade de Agronomia, UFRGS, Brazil
    (1968).
M. Voss, M.Sc. Thesis, Faculdade de Agronomia, UFRGS, Brazil (1981).
L. Waters et al., Hortscience, 15:138-140 (1980).

PLANT FACTORS AFFECTING NODULATION AND SYMBIOTIC

NITROGEN FIXATION IN LEGUMES

P. H. Graham*

Centro Internacional de Agricultura Tropical
AA 6713 Cali- Colombia
Cali, Colombia

INTRODUCTION

Symbiotic N_2 fixation in legumes is the culmination of a complex interaction involving host, Rhizobium, and environment. For fixation to occur, the Rhizobium must find its way into the rhizosphere of an appropriate host, multiply and induce root hair curling be recognized by the plant and gain access to the root system , localize and multiply within root cells, undergo specific structural changes, and receive energy from the plant. Obviously there are many stages in this sequence which can be affected by the host, justifying the contention of Holl and La Rue (1975) that genetic manipulation of the host offers the greatest potential to enhance levels of symbiotic N_2 fixation in legumes.

This review considers those host-controlled traits which influence nodule initiation, development, and function, and shows how some of these traits could be manipulated to obtain enhanced levels of N_2 fixation by plant breeding. It also considers some soil and environmental factors which, through their effect on the host, can limit nodulation and N_2 fixation.

NODULE INITIATION

Table 1 lists plant-controlled characters which can regulate Rhizobium-host recognition and the initiation of infection. Addi-

---

*Present address: Dept. of Soil Science, Univ. of Minnesota, St. Paul, Minn., U.S.A.

TABLE 1. Host Factors Affecting, or Possibly Affecting, Nodule
          Initiation in Legumes

Factor	Legume	Reference
Toxic seed substances	T. subterraneum	Thompson (1960)
	C. pubescens	Bowen (1961)
Root exudate composition	P. sativum	van Egerat (1972)
	G. max	Hubbell and Elkan (1967)
	Cassia	Rao et al. (1973)
Specificity in nodule initiation		
a. Lectin theory of host/strain recognition	T. fragiferum	Dazzo (1980)
b. Cross-inoculation groupings	Various	Various
c. Unusual specificities	T. ambiguum	Hely (1957)
	Lupinus	Lange and Parker (1961)
	G. max	Nangju (1980)
d. Strain selection	G. max	Caldwell and Vest (1968)
		Materon and Vincent (1980)
e. Host environment interactions		
Temperature	P. sativum	Lie et al. (1976)
pH	P. vulgaris	Graham (1982)
	Various	Munns et al. (1977)
f. No nodulation	Various	Various

tional information is provided in the reviews of Broughton (1978)
and Dazzo (1980).

Toxic Seed Substances

     A number of legumes, including Centrosema pubescens (Bowen,
1961), Trifolium subterraneum (Thompson, 1960), Medicago sativa
(Kneur, 1965), and Glycine max (Wahab et al., 1976) possess seed-

coat substances toxic to Rhizobium. When seeds are wet, as during germination, the liberation of these substances can affect Rhizobium survival. It is doubtful, however, that these substances have practical significance. A number of the legumes with seed-coat toxins nodulate normally without specific precautions. Thus, while several authors (Hale, 1976; Jain and Rewari, 1976) have reported better survival of Rhizobium on seeds washed to remove toxic substances, this precaution is probably not worthwhile.

Root Exudates

Virtually all plants excrete a range of sugars, organic acids, and amino acids into the soil immediately adjacent to the root system. Because of these additional nutrients, growth of microorganisms in this region of soil is generally greater than is found in soil devoid of plants. This is generally known as the rhizosphere effect, and the effect is greater for bacteria than for fungi and other organisms. Gram-negative nonsporeforming organisms such as Pseudomonas and Rhizobium are particularly favored, and the number of these organisms in the rhizosphere can exceed the number in soil by a ratio of as much as 100:1. In general, leguminous plants also tend to support larger rhizosphere populations than do nonlegumes. In addition to this nonspecific effect, certain microorganisms can be selectively stimulated. Thus, with legumes, rhizobia in particular benefit, and can achieve concentrations in excess of 10^6/cm root (Bowen, 1961; Marques Pinto et al., 1974). Even finer specificities are possible. Robinson (1967), for example, found that alfalfa stimulated the multiplication of both R. meliloti and R. trifolii, but that with clover, R. trifolii was stimulated much more than R. meliloti.

Rovira (1969) reviewed the chemical nature of root exudates and concluded that such specificity could be a result either of the presence of exotic compounds peculiar to particular plant species or of the balance of the various sugars, amino acids, and organic acids present in the exudates. Two examples support this view: (a) Van Egerat (1972) found that pea roots contain homoserine, a substance specifically favoring the growth of R. leguminosarum. (b) Hubbell and Elkan (1967) studied the root exudates of lines derived from a cross between the soybean cultivars Lee and L9-674, which differed in the ability to nodulate. There was no substance present in exudates from the nodulating isoline which did not also occur in the nonnodulating mutant. However, roots of the nodulating line did have higher levels of protein and reducing sugars than the mutant, and contained less free amino acids. Not all exudates are conducive to Rhizobium development. Cassia fistula, a legume which does not normally nodulate, has been shown to inhibit Rhizobium growth in the rhizosphere (Rao et al., 1973).

Specificity in Infection

Table 1 lists various examples of specificity between Rhizobium and legume in root-hair infection and nodule initiation. It is clear from these interactions, which will be detailed later, that the legume is capable of distinguishing among strains or species of Rhizobium in its root environment. Currently, the only theory which could explain such selectivity is the lectin-recognition hypothesis advanced by Dazzo and Hubbell (1975). According to this theory, sites of nodule initiation on the legume root contain lectins which cross-react with, and bind, carbohydrates on the surface of the appropriate Rhizobium strain.

Evidence in favor of this hypothesis has been obtained mainly with the clover-R. trifolii and soybean-R. japonicum symbiosis. The evidence is as follows. (a) In general, the ability of Rhizobium strains to bind to the roots of particular legumes is correlated with the ability of the strains to nodulate these legumes (Bohlool and Schmidt, 1974). (b) Clover roots bind capsulated cells of R. trifolii (Dazzo and Brill, 1979), FITC-labelled capsular polysaccharide (Dazzo and Brill, 1977), and antibody to clover seed lectin (Dazzo et al., 1978). In each case, binding is most intense in the region of the root-hair zone. (c) Binding is inhibited by the sugar which is the antigenic determinant of the bacterial polysaccharide or the host lectin (Dazzo and Brill, 1977). (d) Intergeneric hybrids of Azotobacter and Rhizobium, which express R. trifolii genes for specific binding with clover lectins (Bishop et al., 1977), also have the ability to bind to clover root hairs (Dazzo and Brill, 1979).

Attractive as this hypothesis is, there are problems still to be resolved. These include the following. (a) Cultivars of P. vulgaris and G. max have been identified which appear to be defective in lectin production but still nodulate (F. A. Bliss, personal communication). (b) Chen and Phillips (1976), Law and Strijdom (1977), and Dazzo and Hubbell (1975b) have noted cases in which binding was observed between certain legume species and heterologous rhizobia. (c) The lectin-recognition hypothesis currently offers no explanation for the competitive differences which exist between Rhizobium strains having essentially similar external polysaccharides. If the lectin hypothesis is valid, it could explain how some environmental factors affect nodule initiation. For example, it has been known for many years that the presence of ammonium or nitrate in the growth media can reduce the number of nodules formed. Dazzo and Brill (1978) showed that the level of immunologically detectable lectin in the host root also declined with ammonium or nitrate addition, perhaps limiting the number of sites available for rhizobial attachment. Interestingly, and in accordance with what happens in the field, low levels of nitrate actually stimulated Rhizobium attachment.

It must be obvious from the points raised above that many more studies of the lectin-recognition hypothesis are still needed. Also necessary are studies on some of the finer specificities which can exist between Rhizobium and host, and cause problems in the field inoculation of some legumes.

The importance of host-Rhizobium specificities was first appreciated by Garman and Didlake (1914), who found six different symbiotic patterns among the different hosts and strains they studied. By 1932, 16 different cross-inoculation groups had been delineated (Fred et al., 1932). Key to the idea of cross-inoculation groups – and to be expected if the mechanism of host-Rhizobium recognition is as simple as proposed by the lectin-recognition hypothesis– – was the concept of reciprocity in nodulation. Thus, Nutman (1956) defined a cross-inoculation group as "... a number of host legumes any one of which will form nodules when inoculated with bacteria from nodules occurring on any other member of the group." Numerous examples of lack of reciprocity in nodulation are available.

Thus: (a) The legume Lupinus cosentinii was introduced accidentally into Western Australia and soon became widely established because of its ability to nodulate and fix N_2 with a range of rhizobia from native legumes (Lange, 1961). By contrast, and while they were supposed to nodulate with the same type of Rhizobium, attempts to introduce other Lupinus species were initially unsuccessful. Plants were generally poorly nodulated and required the provision of specific inoculants before they grew adequately (Lange and Parker, 1961). (b) A similar situation was found in T. ambiguum. Hely (1957) differentiated this species into 2n, 4n, and 6n forms and tested them with rhizobia isolated from T. ambiguum in Australia and New Zealand. Very few induced nodulation. With strains of rhizobia obtained from Turkey, near the center of origin of this plant, nodulation was more common. Nodulation was also obtained when T. repens. was grafted as stock or scion onto T. ambiguum (Hely et al., 1953). (c) The situation with Glycine max is even more complex. This species is native to China with a relatively few lines transported to the United States and used in the breeding of modern soybean cultivars. Now it appears that some Asian cultivars will nodulate freely with native (and presumably cowpea-type) rhizobia (Nangju, 1980), but that strains of rhizobia obtained from China do not nodulate the United States-bred Lee cultivar. This has prompted opposing breeding strategies. In Africa, where there is some doubt about the willingness and ability of farmers to use inoculants, the International Institute of Tropical Agriculture (IITA) and some other scientists are trying to produce cultivars which will nodulate and fix nitrogen with native soil rhizobia. By contrast, and for the United States, where most soils contain large numbers of R. japonicum, and few inoculant strains produce more than 5 to 10% of the nodules formed in the field, Devine and Weber (1977) have pro-

TABLE 2. Competition between Four Strains of R. japonicum for Root
Colonization and Nodulation of Four Cultivars of G. max
(from Materon and Vincent, 1980)

Host cultivar	R_{j2}	Strain representation (%)			
		CB 1809	CC 709	NU 248	NU 150
Lee	−				
Roots (7 days)		41	12	18	29
Nodules		39	12	18	27
Lee × Hardee 36	−				
Roots (7 days)		47	6	18	29
Nodules		42	11	13	32
Hardee	+				
Roots (7 days)		38	7	12	43
Nodules		13	38	8	36
Hardee 31 × Lee	+				
Roots (7 days)		37	4	6	53
Nodules		15	32	7	38

posed that the rj_1 no-nod gene be introduced into United States
varieties so as to limit the number of soil rhizobia able to nodulate these plants.

Even finer specificities have been found in some competition
studies. Thus, Caldwell and Vest (1968) planted the soybean cultivar Peking in soils where strains of the serogroup 110 comprised
60% of the soil population. Despite this, strains belonging to
serogroup 110 formed less than 1% of the nodules. Similarly,
Materon and Vincent (1980) compared the competitive ability of four
R. japonicum strains on lines derived from the cultivar Hardee.
Hardee possesses a gene, rj_2, which permits nodulation but which,
when isolates of the C_1 or 122 serogroups are used, results in
stunted plants unable to fix N_2. Some of the progeny had this gene,
others did not. When cultivars lacking the rj_2 gene were inoculated
with the four strains, strain CB1809 both multiplied aggressively on
the roots and formed a large percent of the nodules. When cultivars
with the rj_2 gene were tested, the number of nodules formed by strain
CB1809 was much lower than would seem likely from the high rhizosphere
population achieved (Table 2).

TABLE 3.   Some Host-Controlled Factors Affecting Nodule Development in Legumes

Factor	Legume	Reference
Bacteroid development	T. pratense	Nutman (1954)
Bacteroid number/ envelope	Lupinus, Ornithopus	Kidby and Goodchild (1966)
Nodule form	Various	Dart (1975)
Nodule number	T. subterraneum	Nutman (1967)
	P. vulgaris	Graham (1973)
Pattern of nodulation	Lupinus sp.	Lange and Parker (1961)
	Sesbania	Dreyfus and Dommergues (1981)
Time to first nodule	T. subterraneum	Nutman (1967)
	T. repens	Gareth Jones and Hardarson (1979)
	Stylosanthes sp.	Graham and Hubbell (1975)
Not specified	Glycine max	Caldwell (1966), Vest (1970)
		Vest and Caldwell (1972)

Host–environment interactions affecting nodule initiation have also been reported. Thus, in Pisum sativum, most cultivars nodulate readily at both 20° and 26°C. However, the cultivar Iran nodulates with all strains at 26°C, but only with strain 310a when a temperature of 20°C is imposed (Lie et al., 1976).

Finally, in this section, we should mention the cultivars of Trifolium pratense, Glycine max, Medicago sativa, Arachis hypogaea, and Pisum sativum which are not nodulated by rhizobia from the appropriate species under a range of temperatures and soil conditions (Aughtry, 1948; Nutman, 1949; Williams and Lynch, 1954; Gorbert and Burton, 1980). In red clover, the no-nodulation characteristic is the result of a single recessive gene which operates in conjunction with a cytoplasmic factor (Nutman, 1949). In this trait root hairs are formed, and do curl, but Rhizobium strains appear incapable of penetrating them. In peanut, Nambiar et al. (1982) identified two recessive genes controlling the lack of nodulation, but they found that in some crosses with plants having one or other of the genes, a few very large nodules were formed. The mechanism involved has not been explained.

TABLE 4. Some Host-Controlled Factors Affecting Nodule Function
          in Legumes

Factor	Legume	Reference
Bacteroid function	P. sativum	Holl (1973)
	M. sativa	Viands et al. (1979)
Delayed senescence	G. max	Abu Shakra et al. (1978)
Effectiveness subgroups	Various	Vincent (1974)
		Date and Norris (1979)
Energy requirement for fixation	Lupinus, Vigna	Layzell et al. (1980)
Energy partitioning	Various	Graham and Halliday (1977)
Form of N exported	Various	Pate (1980)
Hydrogenase regulation	P. sativum	Dixon (1972)
Leghaemoglobin synthesis	Lupinus, Ornithopus	Dilworth (1969)
Maturity characteristics	G. max	Hardy et al. (1973)
Not specified	T. subterraneum	Gibson (1964)
	T. pratense	Nutman (1968)

Lack of nodulation is surprisingly common in pea cultivars.
Lie (1971) showed that the cultivar Afganistan did not nodulate
with any of 20 strains of R. leguminosarum, and later (Lie et al.,
1978) reported a number of other cultivars, obtained from the center
of origin of this crop in the Middle East, which also did not nodulate.

NODULE DEVELOPMENT

Table 4 lists a number of host-controlled traits which can
affect nodule development. The majority of these are not of economic
significance so I will emphasize only nodule form, nodule number,
pattern of nodulation, time to first nodule appearance, and some
genetic abnormalities.

Dart (1975) distinguished three basic nodule types: (a) elon-
gate and cylindrical, with apical meristematic activity, as in clover
and medic; (b) spherical, with several discrete meristematic foci,
as occur in soybean and bean; and (c) collar nodules, as in lupin,
where the nodule extends about the root. Nodules with apical meri-

stematic activity may later develop a branched or even corolloid structure through dichotomy of the meristem. Nodules in some legumes are perennial with a banded appearance resulting from altnerating periods of growth and no-growth.

Legume species vary markedly in the number of root hairs infected by the appropriate Rhizobium strain. Neptunia, an aquatic species, has no root hairs, while in species of Trifolium, the number of infected hairs in 11-day-old plants can range from 2 to 80 (Nutman, 1962). The proportion of these infected hairs which give rise to nodules can also vary. Thus, Roughley et al. (1970) found that lines bred for sparse or abundant nodulation could have similar numbers of infected root hairs. This and other evidence have led to the belief that nodules can only form at specific sites on the root, and thus that nodule number is essentially a host-controlled trait. Wipf and Cooper (1940) proposed that naturally occurring disomatic cells in the root cortex were necessary for nodule initiation, whereas Oinuma (1948) has reported that disomatic cells result from the plant hormones rhizobia excrete. Be that as it may, host- and strain-controlled differences in nodule number have now been found in a number of species, including Trifolium subterraneum (Nutman, 1967) and Phaseolus vulgaris (Graham, 1973). Nutman (1967) reported nodule number in 15 cultivars of subterranean clover to vary from 18.3 to 108.1 per plant. The tendency toward high nodule number was dominant but probably controlled by more than one gene. High nodule number was correlated with lateral root number and inversely related to nodule size, making it a somewhat unreliable criterion on which to base inoculant success. For this reason, it is only used in our laboratory when we believe that the treatments imposed could have affected inoculant survival.

The pattern of nodulation on legumes, while affected by such factors as temperature (Munns et al., 1977b), planting density (Graham and Rosas, 1978), and delayed inoculation (Dart and Pate, 1959), can also vary with the host. Thus, Lange and Parker (1960) found L. consentinii predominantly tap-root nodulated, whereas in L. mutabilis, most nodules were found on the crown and on lateral roots. In Arachis and many Stylosanthes species, nodules are found only at sites of lateral-root emergence, whereas in Sesbania rostrata, stem nodules are found (Dreyfus and Dommergues, 1980) and are claimed to continue fixing even in inundated plants or plants supplied combined N.

Host-controlled differences in time to the appearance of first nodule have been found in a number of legumes, and are commonly also strain dependent. Thus, Nutman (1967) found that 10 cultivars of subterranean clover ranged in time to first nodule appearance from 14.8 to 18.2 days when inoculated with strain SU 297, and from 14.9 to 20.7 days when strain SU 220 was used. A similar pattern has been observed in Stylosanthes species (Graham and Hubbell, 1975)

and white clover (Gareth Jones and Hardarson, 1979). Use of culti-
var-strain combinations which nodulate rapidly could be one means of
limiting competition from native soil rhizobia. Thus, Gibson (1962)
reported 50% of the native soil rhizobia slow to nodulate with the
subclover cultivar Woogenellup. In pot trials with soils which contained 10^4 soil rhizobia per gram, some inoculant strains still produced more than 80% of the nodules. Strain TAI, also slow to nodulate Woogenellup, was, however, seriously disadvantaged. A similar
situation has been reported by Gareth Jones and Hardarson (1979).
These authors added mixtures of strains to three clover cultivars in
soil and typed the nodules which resulted. With cultivar S100,
strain 75^{str} formed almost 100% of the nodules; it formed 25.1% on
S strain 184 and 28.4% on the cultivar Pajbjerg smalblackt. Preference for strain 75 was heritable (Hardarson and Gareth Jones, 1979a).
In a subsequent paper (Hardarson and Gareth Jones, 1979b), host-
strain selection was evaluated as a function of temperature. In
these two papers, there was a close relationship between host preference for strain 75^{str} and the time each strain took to nodulate.
When strain 75^{str} was the faster to nodulate, it occupied most
nodules; when strain 33^{sp} was faster, it did.

A number of cultivars with defects in nodule development have
been identified and, in some cases, characterized. Nutman (1954a,
1954b, 1957) reported two recessive conditions in red clover, termed
i_1 and ie. Bergersen and Nutman (1957) determined that the i_1 condition was a result of the failure of the rhizobia in the host cells
to produce bacteroids, whereas the ie condition was the result of a
tumor-like growth within the nodule, again without bacteroid formation. Three ineffective conditions exist in soybean (Caldwell,
1966; Vest, 1970; Vest and Caldwell, 1972) and are recognized by the
symbols rj_2, rj_3, and rj_4. Caldwell (1966) found that the cultivar
Hardee, when inoculated with isolates of the C1 or 122 serogroup,
was stunted and N-deficient. The rj_3 gene is also found in the
cultivar Hardee but is only expressed when strain 33 is used as inoculant. As with rj_2, this gene is dominant, but it can be distinguished from rj_2. The rj_4 gene is expressed in the cultivar Hill
inoculated with strain 64. The source of this gene is the cultivar
Dunfield.

NODULE FUNCTION

Table 4 lists some host-controlled traits affecting nodule
function. Three of these traits are the result of loss mutations
at specific loci. Thus, Viands et al. (1979) described a trait in
Medicago sativa in which nodule development was normal, or even enhanced, but starch accumulated in the nodule and N_2 fixation was
limited. The ineffective association of T. subterraneum cultivar
Northam First Early with strain NA 30 has not been characterized,
but both host and Rhizobium symbiose well with other partners. The

sym[3] condition detailed by Holl and La Rue (1975) will be mentioned
in more detail later.

Effectiveness Subgroups

Subgroups which vary in their ability to fix N_2 with particular
Rhizobium strains exist in many of the so-called cross-inoculation
groups.

In the clover group, for example, at least six different inoculant cultures are needed to cater to the differing symbiotic
responses of species in this genus (Vincent, 1974). Thus, strains
of R. trifolii effective with T. repens and T. pratense are usually
ineffective on T. subterraneum (Vincent, 1954). This has been a
problem in Australia, where white clover has established widely in
the native pastures, but is often replaced with subterranean clover.
The difficulty in the nodulation of T. ambiguum has already been
mentioned. Even more complicated is the situation with African
clover species. Norris (1959) examined eight species of Trifolium
from equatorial Africa. Some nodulated with strains from Vicia and
Pisum but did not nodulate, or were ineffective, with rhizobia from
European clovers. In a more detailed study, Norris and t'Mannetje
(1964) divided nine species of African clovers into at least four
different groups, each of which required a different inoculant.

The situation in the genus Stylosanthes is even more complex.
For a number of years in Australia, the only strain recommended for
use with Stylosanthes species was CB 756 (Date, 1969). However,
when 336 accessions from 15 species of Stylosanthes were tested with
this strain, only 224 gave an effective response (Date and Norris,
1979). The accessions giving an ineffective response with CB 756
included 67 of 143 accessions of S. guianensis and 31 of 41 accessions of S. hamata. When these accessions were further tested with
22 strains of Rhizobium isolated from various Stylosanthes species,
18 accessions gave ineffective responses with all strains. An
effective response pattern in Stylosanthes has been related to cultivar habitat (Date et al., 1979) and, in S. guianensis, with seed
isozyme patterns (Robinson et al., 1976).

Energy Requirement for Fixation

Differences between cultivars of the same species in ability
to fix N_2 in symbiosis with Rhizobium have now been demonstrated in
a range of legumes including clover (Gibson and Brockwell, 1968;
Mytton, 1975) soybean (Hardy et al., 1973), beans (Graham and Halliday, 1977; Graham, 1981; Rennie and Kemp, 1981), Medicago (Gibson,
1962; Seetin and Barnes, 1977), cowpea (Zary et al., 1978), Vicia
(El Sherbeeny et al., 1977), Pisum (Holl and La Rue, 1975), and
Desmodium (Hutton and Coote, 1972; Imrie, 1975). In other studies
(Sen and Weaver, 1980), differences in SNA have also been demonstrated between peanut and cowpea.

TABLE 5. Economy of Carbon and Nitrogen in Nodulated Roots of
          Selected Legumes (from Minchin et al., 1981)

Attribute	Pea	White lupin	Cowpea
N_2 fixed (mg/plant/day)	3.0	10.6	14-50
C exported from nodule (mg/plant/day)	5.9	24.2	21.1
C for nodule development (mg/plant/day)	4.9	39.9	35.5
C loss, respiration (mg/plant/day)	18.0	114.4	113.3
Total C consumed (mg/plant/day)	28.7	178.5	169.6
C respired:N_2 fixed (mg/mg)	6.0	10.7	6.3

While these differences can be compounded by host-strain interactions (Mytton et al., 1977; Minchin et al., 1978) and climatic
differences (Graham, 1981; Rennie and Kemp, 1981), they appear to
stem mainly from differences in the supply of carbohydrate.

While, in theory, N_2 fixation is an exergonic reaction (Bayliss,
1956), it has now been clearly established that the process requires
a source of photosynthate. In part, this is because the bacteroids
derived from many strains of Rhizobium are inefficient and waste as
much as 25% of the energy available to them in the reduction and
evolution of hydrogen. Because of this hydrogen evolution, Minchin
et al. (1981) give the equation for N_2 fixation as:

$$N_2 + 4 \text{ NADH} + 6 \text{ H}^+ + 16 \text{ ATP} \rightarrow 2NH_4^+ + H_2 + 4 \text{ NAD} + ADP$$

Energy is also required for the development of nodules and their
maintenance. In Table 5 (from Minchin et al., 1981) are data for
peas, lupin, and cowpea. In other studies reviewed by these authors,
3.0 to 6.9 mg C is respired for each mg of N_2 fixed, with 12.8 to
28.2% of the total photosynthate derived by the plant used in nodule
function and N_2 fixation. Because of the photosynthate required for
N_2 fixation, it is widely assumed that plants dependent on N_2 will
have less photosynthate available for growth than those supplied
nitrate (Schubert and Ryle, 1980). However, Neves (1982) points out
that the theoretical cost of using nitrate is similar to that of N_2
fixation. Whole plant studies undertaken to date have given conflicting results, depending on the level of nitrate applied.

Because of the photosynthetic requirement for N_2 fixation,
manipulations which impair photosynthesis will reduce N_2 fixation,

TABLE 6. Effect of CO_2 Air Enrichment on Some Parameters of Yield and N_2 Fixation in Soybeans (from Hardy and Havelka, 1976)

Parameter	Air	CO_2-enriched air
Yield (g/plant)	8.8	15.7
Biological yield (g/plant)	20.6	30.7
N_2 fixed (g/plant)	0.17	0.84
(kg/ha)	76	427
% Plant N from soil	73	16
Maximum nodule fresh wt (g/plant)	3.5	9.0
Maximum SNA (μmole C_2H_4/mg nodule fresh wt/day)	350	550

while those which enhance photosynthesis will increase it (Lawrie and Wheeler, 1973; Lawn and Brun, 1974; Minchin and Pate, 1974; Hardy and Havelka, 1976; Herridge and Pate, 1977; Bethlenfalvay et al., 1978). The short-term effects of such treatments are principally on fixation efficiency (N_2 fixation/g nodule tissue/unit time), but where long term treatments are applied, changes in nodule mass can also occur. Several examples will illustrate these points.

Hardy and Havelka (1976) used CO_2 enrichment of the air passing through soybean canopies to enhance photosynthesis under field conditions. Enriched plots received 800 to 1200 ppm CO_2 from 38 days maturity, and control plots received air alone. Results from this experiment are shown in Table 6. Plant weight was markedly enhanced in this trial with both nodule fresh weight and average specific nodule activity doubled. As a result, the CO_2-enriched plants fixed 427 kg N/ha, almost 6 fold the fixation found in the control plants.

Graham and Rosas (1978) studied the effect of planting density on N_2 fixation in P. vulgaris. The three cultivars studied varied strikingly in response to planting density. The climbing cultivar showed a sharp peak in fixation at a relatively low density. Below the peak density, extra light for the lower leaves was probably balanced by higher temperatures, and consequently nodule respiration, in the exposed soil. At higher densities, competition for light between plants would have reduced photosynthesis. A heavily branched, sprawling bush cultivar showed no such sharp peak, presumably because of its ability to cover the ground even at low plant densities.

Lawn and Brun (1974) used both shading and supplemental light-ing. In their trial, pod removal (to limit the competition for pho-tosynthate between developing pods and nodules) markedly stimulated fixation. Other workers have had difficulty in consistently achiev-ing this result, perhaps because of hormonal changes associated with pod abscission.

Source of the photosynthate for nodules is another important factor. Waters et al. (1980), using P. vulgaris, showed that 85% of the $^{14}CO_2$ absorbed by the leaf at node 4 passed to root and nodules, but that fed to leaves on higher nodes, the radioactivity tended to remain in the above-ground parts. Leaves at the lower nodes are subject to shading and tend to decline in efficiency after canopy closure (Tanaka and Fujita, 1979).

A number of host traits can each affect carbohydrate supply to nodules. The best documented among these is the time varieties take to flower and mature. Thus, Hardy et al. (1973) demonstrated that early flowering soybean lines tended to fix less N_2 than those from the later maturity groups. This is presumably again because of com-petition between developing pods and nodules. Hardy and his co-workers have hypothesized that a 9 day delay in flowering would al-low a doubling in seasonal rates of N_2 fixation, and N_2 fixation in beans has been significantly enhanced by a photoperiod-induced de-lay in flowering (J. Day and P. H. Graham, unpublished data). Un-fortunately, extending the growth cycle of legume cultivars in the tropics, where many grain legumes are already subject to water stress at flowering, could be difficult.

Leaf-area duration may also be important. Wynne et al. (1982), in studies with peanuts, found that 70 to 75% of the variation in nodulation and N_2 fixation found in eight peanut cultivars could be attributed to difference in leaf-area duration.

Ability to partition carbohydrate between root, shoot, and nodules must also be critical to nodule function. Thus, Adams et al. (1978) found bean cultivars highly variable in carbohydrate accumu-lation and remobilization, with the lowest yielding cultivars those that accumulated carbohydrate in stem and root but did not remo-bilize it during seed development. Accumulation of carbohydrate in roots is characteristic of several of the ineffective traits listed in Table 5, the sym 3 condition detailed by Holl and La Rue (1975) is actually capable of N_2 fixation when supplied succinate as an energy source. Male sterile soybeans have been shown to accumulate up to five times normal carbohydrate levels in the root with only limited benefit to fixation (Wilson et al., 1978). Among commercial bean cultivars, Graham and Halliday (1977) identified some lines es-sentially similar in total root system carbohydrate but differing by 5 to 100 fold in the carbohydrate recovered from nodules.

ADDITIONAL TRAITS AFFECTING NODULATION
AND N₂ FIXATION AND POSSIBLY UNDER HOST CONTROL

To this point, we have considered only traits which are known
to be under host control. In this section, I would like to be specu-
lative and to consider traits which definitely affect nodulation or
N₂ fixation, but for which evidence of host plant control is limited.

Inundation and Drought

Water potentials near to field capacity are usually considered
necessary for nodulation and N₂ fixation. Water deficits of only
2.5 bars can cause significant reduction in SNA, and at −15 to −20
bars, fixation is essentially eliminated. Recovery from drought
can take a considerable time, and with prolonged drought, loss of
fixation may be essentially complete.

Inundation also causes reduced fixation, with many nodules shed
after only 1 to 2 days inundation. Again, recovery from inundation
can take considerable time, especially when inundation occurs soon
after initial nodulation. Recovery from inundation can also be slow,
though cases have been reported where rapid renodulation and nodule
development posed a very severe strain on the carbohydrate reserves
of the plant, and reduced yield. Legumes vary in their tolerance
to the extremes of water shortage and excess, with desert plants
such as Prosopis and Dalea tolerant of dry conditions and plants
such as Neptunia and Aeschynomene tolerant of inundation. Among
cultivated legumes, cowpea and tepary bean are most used in dried
environments. In P. vulgaris, some cultivars tolerate inundation
well, with yield losses of only 17% following prolonged flooding:
in others, yield losses of up to 70% have been reported. To my
knowledge, none of the cultivar comparisons for drought or flooding
tolerance has considered N₂ fixation. At CIAT, we have some indications though that some cultivars recover N₂ fixation potential
after flooding more rapidly than others. More detailed studies are
needed.

Tolerance of Low Soil Phosphorus Levels

Phosphorus is essential for legume growth and N₂ fixation. In
P-deficient soils, nodule mass, N₂ fixation per plant, and SNA can
all be limited (Graham and Rosas, 1979). Plants dependent on N₂
fixation can require up to 50% more P than those supplied fertilizer
N (Cassmann et al., 1981). At the same time, not all legumes require the same level of P, nor do they respond the same to P deficiency.

Andrew and Robins (1969) compared the P requirements of nine
tropical and one temperate pasture legumes in two soils deficient
in P. They found Glycine javanica and Desmodium intortum highly re-

sponsive to applied P, whereas <u>Stylosanthes humilis</u> gave excellent
dry matter production even at low levels of applied P. Because of
rising fertilizer prices, a number of studies have attempted to identify such differences in P requirements among cultivars of the same
species. Thus, in <u>P. vulgaris</u>, 54 lines screened by Whitaker et
al. (1976) varied in dry matter production per mg of P supplied from
380 to 671 mg. Under P stress, net photosynthesis per unit P varied
significantly between cultivars, suggesting that the differences observed were not only a result of P uptake. In recent years at CIAT,
there have been a number of experiments to select cultivars of <u>P.
vulgaris</u> both efficient in P use and responsive to applied P. Cultivars are grown at 50 and 300 kg P_2O_5/ha, and those cultivars are
selected which give good yields at the lower figure but which are
also responsive to P.

Studies with white clover have also been undertaken using the
two P levels approach. Again the results suggested that there were
differences between lines in respect to P nutrition, but that these
differences were probably quite small (Caradus et al., 1980). Several different mechanisms have been invoked to explain differences
between species and/or cultivars in P requirement; these could have
implications for N_2 fixation. These incude: inherent differences
in ability to absorb P; differences in critical P content; variable
response to mycorrhiza; and partitioning differences within the
plant. I will not dwell on these but will give some examples and
will show how these could affect N_2 fixation. Thus:

(a)   Andrew and Robins (1969a) found the critical percentage
      P required by 10 pasture legumes to vary from 0.16 to
      0.25%. Though <u>S. humilis</u> had a critical percent for P
      of only 0.17%, it was able to accumulate appreciable N in
      the low N soil used, presumably by N_2 fixation. By con-
      trast, <u>G. javanica</u>, with a critical percent for P of 0.23%,
      only accumulated significant N at higher levels of P fer-
      tilization (Andrew and Robins, 1969b).

(b)   Yost and Fox (1979) compared four legume species for de-
      pendence on mycorrhiza. When soil was fumigated to sup-
      press mycorrhiza, <u>Glycine max</u> and <u>Vigna unguiculata</u>
      achieved P uptake similar to the unfumigated controls at
      0.1 to 0.2 µg P/ml of soil solution, whereas <u>S. hamata</u> re-
      quired 1.6 µg P/ml. The authors concluded that in low P
      soil conditions, <u>S. hamata</u> would depend heavily on mycor-
      rhizal infection. Correlations between levels of mycor-
      rhizal infection and N_2 fixation have been shown for a
      number of legumes (Crush, 1974; Draft and El-Giahmi,
      1974; Carling et al., 1978). Unfortunately, there ap-
      pears little difference between cultivars in ability to
      associate with mycorrhizae (Crush and Caradus, 1980).

(c) At CIAT, we have looked at the difference in P uptake and
     N_2 fixation of cultivars differing in tolerance to low
     soil P. At 16 ppm P, N accumulation per plant was greater
     in the inefficient cultivar. However, at 4 ppm P, N ac-
     cumulation by the P-tolerant cultivar was as much as twice
     that of the inefficient cultivar. We are currently look-
     ing at the inheritance of P tolerance in P. vulgaris and
     how this relates to N_2 fixation.

Soil Acidity and Nodulation

Legumes show large differences in tolerance of soil acidity.
Unfortunately, most of the studies to date have failed to separate
host and Rhizobium effects or have ignored N source completely.
With differences in the acid tolerance of Rhizobium strains increasingly evident (Keyser et al., 1979; Graham et al., 1982), it
is obvious that studies on acid soil tolerance must consider not
only both host and Rhizobium, but also distinguish effects on
nodule initiation from those which limit fixation. Differences between species in nodulation and N_2 fixation at low pH have been
clearly demonstrated. Thus, Munns et al. (1977) grew a range of
legumes on a Hawaiian oxisol of pH 4.8, and they measured the
effects of liming on yield and N_2 fixation. V. unguiculata, Arachis
hypogaea, Glycine max, and Stylosanthes spp. proved acid tolerant,
with yields increased less than 30% with limiting and maximized at
1 ton $CaCO_3$/ha applied, whereas Leucaena leucocephala, Coronilla
varia, Medicago sativa, and Phaseolus vulgaris proved acid sensitive,
with yields increased more than 5 fold with liming and many tons of
lime required for maximum yield. In these trials, N_2 fixation by
Stylosanthes sp. was little affected by acid pH, whereas Leucaena,
although well nodulated at low pH, responded to lime, suggesting
that nodule function was acid sensitive.

While intraspecific (cultivar) differences in tolerance to acid
soil conditions have been shown for a number of legumes, including
bean and cowpea (Spain, 1976), most such studies have not considered
possible effects on N_2 fixation. The work of Bouton et al. is an
exception. These authors selected within the Florida 66 variety of
Medicago sativa for vigor in acid (pH 4.4) or limed soils. When the
selected lines were grown at pH levels from 4.4 to 6.1, the acid-
selected germplasm showed greater nodule fresh weight and N_2 fixation at pH 4.4 than at pH 6.1, whereas the reverse was true for materials selected at higher pH values. The acid-selected germplasm
tended to outyield that selected under limed conditions. At CIAT,
we have evaluated approximately 70 lines of P. vulgaris for growth
and N_2 fixation at pH 4.5, and we have found considerable variation.
We are now determining the effect of added Al and Mn on pH-tolerant
germplasm.

IMPLICATIONS FOR BREEDING

Holl and La Rue (1975) suggested that a minimum of 10 host genes controlled nodulation and N_2 fixation in legumes. From points made in this presentation, and as they themselves suggested, this number is likely to be extremely conservative. With so many genes involved, it is not really surprising that we continue to identify cultivars defective in some trait limiting nodulation or N_2 fixation. What is important is that plant breeders and microbiologists cooperate to control this problem, with no lines distributed before they have been evaluated for symbiotic potential. At CIAT, all lines produced by the plant breeders and being considered for release to national programs must undergo a series of tests to define resistance to diseases, yield potential, cooking quality, etc. Included in these tests are both a field and a glasshouse evaluation of N_2 fixation. For the glasshouse evaluation, nodulated plants are grown on quartz sand without N, but they are supplied all other nutrients. Thus, they are completely dependent on fixed N for growth and final yield. To date, yield per plant under these no N conditions have ranged from 2 to 22 g per plant. In the field study, plants grown on N-deficient soil are evaluated by nodule dry weight and N_2 (C_2H_2) reduction at flowering and for final yield. Lines weak in N_2 fixation are discarded.

On the positive side, it is becoming increasingly evident that host-controlled differences, such as those reported in this paper, can be used in producing agronomically acceptable cultivars active in N_2 fixation. With the prices of fertilizer N already high and likely to increase sharply in the next five years, an increasing number of laboratories are looking to produce cultivars active in N_2 fixation (Graham, 1981; McFerson et al., 1982; Wynne et al., 1982). Initial results have been very promising. Thus, McFerson et al. (1982) used a recurrent selection procedure to produce lines similar in appearance and grain type to the poor N_2 fixer, Sanilac, but having fixation similar to the cultivar P498, which is active in N_2 fixation. Transgressive segregation was found with some lines better in N_2 fixation and yield than either parent. At CIAT, a number of lines active in N_2 fixation with selected Rhizobium strains have also been identified. With the potential to enhance N_2 fixation admitted, there are still many problems to be faced.

Early studies on the inheritance of N_2 fixation looked only at F_2 or S_1 populations and used destructive methods to evaluate variation in nodulation or N_2 fixation. To be effective, such early generation screening must allow production of seed for later multiplication. Various solutions to this problem are now being attempted. In our laboratory, as already mentioned, we use yield

under low N as an indicator of fixation. Mahon and Salminen (1980)
use a modified acetylene-reduction technique in which acetylene is
injected directly into soil at the foot of field-grown plants; analyses are then made of the soil atmosphere after varying periods of
incubation. For chickpea, Rupela and Dart (1982) have developed a
method for the evaluation of nodules, following which selected plants
are repotted with a high degree of recovery. More study of these
different techniques and their reliability is urgently needed.

Another problem is that of the strain or strains to be used.
With soybean, for example, most soils in the United States contain
large numbers of R. japonicum, and inoculant strains are normally
responsible for less than 10% of the nodules. Breeding for enhanced
N$_2$ fixation with standard inoculant strains may thus have little
value, if these strains are incapable of producing a significant
number of nodules. Two alternate approaches seem justified. (a)
Breeding host cultivars which are not specific in their response
and fix N$_2$ with a wide range of native soil isolates. In soybean,
use of the Orba or Mandarin varieties in breeding would be one example of this approach. Kvien et al. (1981) have also examined a
range of soybean cultivars for compatibility with the predominant
123 serogroup in Minnesotan soils. (b) The alternative approach,
suggested by Devine and Weber (1977) and possibly the results of
Hardarson and Gareth-Jones (1979a), would be to breed for an extremely specific reaction such that most of the native soil rhizobia
either do not nodulate the legume in question or do so so slowly
that they would not be able to compete with the inoculant strains.
This has already happened accidentally on several occasions, for
example, the stylo cultivar 1022 evaluated by Souto et al. (1972).

Finally, when considering methodologies, there is question of
combining enhanced rates of N$_2$ fixation with other needed agronomic
and disease-resistance traits. It is not much use producing cultivars which actively fix N$_2$ if they all die from rust or anthracnose
when grown in the field. The problem here is that both N$_2$ fixation
and resistance to a number of diseases are likely to have polygenic
inheritance. We must combine disease resistance and N$_2$ fixation at
an early stage in the breeding program and attempt to pyramid levels
of N$_2$ fixation and disease resistance. This demands active col-
laboration between microbiologist, plant breeder, and pathologist.

In summary, while many problems remain to be overcome, the possibility of developing varieties that are both active in N$_2$ fixation and are agronomically satisfactory seems feasible. Crops for
which breeding initiatives should be undertaken include soybean,
bean, pigeon pea, peanut, and a number of lesser legumes. I look
forward both to rapid progress and many new challenges in the future.

REFERENCES

S. A. Abu-Shakra, D. A. Phillips, and R. C. Huffaker, Science, 199: 973–975 (1978).

M. W. Adams, J. V. Wiermsa, and J. Salazar, Crop Sci., 18:155–157 (1978).

C. S. Andrew and M. F. Robins, Aust. J. Agric. Res., 20:665–674 (1969a).

C. S. Andrew and M. F. Robins, Aust. J. Agric. Res., 20:675–685 (1969b).

J. D. Aughtry, Cornell Univ. Agric. Exp. Sta. Memoir 280, Ithaca, New York (1948).

N. S. Bayliss, Aust. J. Biol. Sci., 9:364–370 (1956).

F. J. Bergersen and P. S. Nutman, Heredity, 11:175–184 (1957).

G. J. Bethlenfalvay, S. S. Abut-Shakra, K. Fishbeck, and D. A. Phillips, Physiol. Plant, 43:31–34 (1978).

P. E. Bishop, F. B. Dazzo, E. R. Appelbaum, R. J. Maier, and W. J. Brill, Science, 198:938–940 (1977).

B. B. Bohlool and E. L. Schmidt, Science, 185:269–271 (1974).

J. H. Bouton, M. E. Summer, and J. E. Giddens, Plant Soil, 60:205–211 (1981).

G. D. Bowen, Plant Soil, 15:155–165 (1961).

W. J. Broughton, J. Appl. Bacteriol., 45:165–194 (1978).

B. E. Caldwell, Crop Sci., 6:427–428 (1966).

B. E. Caldwell and G. Vest, Crop Sci., 8:680–682 (1968).

J. R. Caradus, J. Dunlop, and W. M. Williams, N. Z. J. Agric. Res., 23:211–217 (1980).

D. E. Carling, W. G. Riehle, M. F. Brown, and D. R. Johnson, Phytopathology, 68:1590–1596 (1978).

K. G. Cassmann, A. S. Whitney, and R. L. Fox, Agron. J., 73:17–22 (1981).

P. C. Chen and D. A. Phillips, Plant Physiol., 59:440–442 (1977).

J. R. Crush, New Phytol., 73:745–754 (1974).

J. R. Crush and J. R. Caradus, N. Z. J. Agric. Res., 23:233–237 (1980).

M. J. Daft and A. A. El-Giahmi, New Phytol., 73:1139–1147 (1974).

P. J. Dart, in: The Development and Function of Roots (J. G. Torrey and D. T. Clarkson, eds.), Academic Press, London, pp. 467–506 (1975).

P. J. Dart and J. S. Pate, Aust. J. Biol. Sci., 12:427–444 (1959).

R. A. Date, J. Aust. Inst. Agric. Sci., 35:27–37 (1969).

R. A. Date, R. L. Burt, and W. T. Williams, Agro-Ecosystems, 5:57–67 (1979).

R. A. Date and D. O. Norris, Aust. J. Agric. Res., 30:85–104 (1979).

F. B. Dazzo, in: Advances in Legume Science (R. J. Summerfield and A. H. Bunting, eds.), Royal Botanic Gardens, Kew., pp. 49–59 (1980).

F. B. Dazzo and W. J. Brill, Appl. Environ. Microbiol., 33:132–136 (1977).

F. B. Dazzo and W. J. Brill, Plant Physiol., 62:18–21 (1978).

F. B. Dazzo and W. J. Brill, J. Bacteriol., 137:1362-1373 (1979).

F. B. Dazzo and D. H. Hubbell, Appl. Microbiol., 30:1017-1033 (1975).

F. B. Dazzo, W. E. Yanke, and W. J. Brill, Biochim. Biophys. Acta, 539:276 (1978).

T. E. Devine and D. F. Weber, Euphytica, 26:527-535 (1977).

M. J. Dilworth, Biochim Biophys. Acta, 184:432-441 (1969).

R. O. Dixon, Arch. Mikrobiol., 8:193-201 (1972).

B. L. Dreyfus and Y. R. Dommergues, in: Current Perspectives in Nitrogen Fixation (A. H. Gibson and W. E. Newton, eds.), Australian Academy of Science, Canberra (1981).

M. H. El Serbeeny, D. A. Lawes, and L. R. Mytton, Euphytica, 26: 377-383 (1977).

E. B. Fred, I. L. Baldwin, and E. L. McCoy, Root Nodule Bacteria and Leguminous Plants, University of Wisconsin Press, Madison, Wisconsin (1932).

D. Gareth-Jones and G. Hardarson, Ann. Appl. Biol., 92:221-228 (1979).

H. Garman and M. Didlake, Kentucky Agric. Exp. Sta. Bull., 184 (1914).

A. H. Gibson, Aust. J. Agric. Res., 13:388-399 (1962).

A. H. Gibson, Aust. J. Agric. Res., 15:37-49 (1964).

A. H. Gibson, Aust. J. Agric. Res., 19:907-918 (1968).

A. H. Gibson and J. Brockwell, Aust. J. Agric. Res., 19:891-905 (1968).

D. W. Gorbert and J. C. Burton, Crop Sci., 19:727-728 (1979).

P. H. Graham, in: Genes, Enzymes and Populations (A. M. Srb, ed.), Plenum Press, New York, pp. 321-330 (1973).

P. H. Graham, Field Crops Res., 4:93-112 (1981).

P. H. Graham and J. Halliday, in: Exploiting the Legume-Rhizobium Symbiosis in Tropical Agriculture (J. M. Vincent et al., ed.), Univ. of Hawaii Coll. Trop. Agric. Misc. Publ. 145, pp. 313-334 (1977).

P. H. Graham and D. H. Hubbell, Florida Agric. Exp. Sta. J. Ser., 5439, pp. 9-21 (1975).

P. H. Graham and J. C. Rosas, J. Agric. Sci. (Cambridge), 90:19-29 (1978).

P. H. Graham and J. C. Rosas, Agron. J., 71:925-926 (1979).

P. H. Graham, S. E. Viteri, F. Mackie, A. T. Vargas, and A. Palacios, Field Crops Res., in press.

C. N. Hale, Proc. N. Z. Grassland Assoc., 38:182-186 (1976).

G. Hardarson and D. Gareth Jones, Ann. Appl. Biol., 92:329-333 (1979a).

G. Hardarson and D. Gareth Jones, Ann. Appl. Biol., 92:229-236 (1979b).

R. W. F. Hardy, R. C. Burns, and R. D. Holstein, Soil Biol. Biochem., 4:47-82 (1973).

R. W. F. Hardy and U. D. Havelka, in: Symbiotic Nitrogen Fixation in Plants (P. S. Nutman, ed.), Cambridge Univ. Press, London, pp. 421-439 (1976).

F. W. Hely, Aust. J. Biol. Sci., 10:1–16 (1957).

F. W. Hely, C. Bonnier, and P. Manil, Nature (London), 171:884–885 (1953).

D. R. Herridge and J. S. Pate, Plant Physiol., 60:759–764 (1977).

F. B. Holl, Can. J. Genet. Cytol., 15:659 (1973).

F. B. Holl and T. A. La Rue, in: International Symposium on Nitrogen Fixation, Vol. 2, Univ. of Washington Press, Pullman, Washington, pp. 391–399 (1975).

D. H. Hubbell and G. H. Elkan, Phytochemistry, 6:321–328 (1967).

E. M. Hutton and J. N. Coote, J. Aust. Inst. Agric. Sci., 38:68–69 (1972).

B. C. Imrie, Euphytica, 24:625–631 (1975).

M. K. Jain and R. B. Rewari, Zent. Bakt. Parasitenk. Abt. 2, 131, 163–169 (1976).

H. H. Keyser, D. N. Munns, and J. S. Hohenberg, Soil Sci. Soc. Am. J., 43:719–722 (1979).

D. K. Kidby and D. J. Goodchild, J. Gen. Microbiol., 45:147–150 (1966).

C. E. Kneur, Cited in Plant Breeding Abstr., 37:815 (1967).

C. S. Kvien, G. E. Ham, and J. W. Lambert, Agron. J., 73:900–905 (1981).

R. T. Lange, J. Gen. Microbiol., 26:351–359 (1961).

R. T. Lange and C. A. Parker, Plant Soil, 13:137–146 (1960).

R. T. Lange and C. A. Parker, Nature (London), 187:178–179 (1961).

I. J. Law and B. W. Strijdom, Soil Biol. Biochem., 9:79–84 (1977).

R. J. Lawn and W. A. Brun, Crop Sci., 14:11–16 (1974).

A. C. Lawrie and C. T. Wheeler, New Phytol., 72:1341–1348 (1973).

D. B. Layzell, R. M. Rainbrid, C. A. Atkins, and J. S. Pate, Plant Physiol., 64:88–89 (1979).

T. A. Lie, Plant Soil, 34:751–752 (1971).

T. A. Lie, Ann. Appl. Biol., 88:462–465 (1978).

T. A. Lie, D. Hille, R. Lambers, and A. Houwers, in: Symbiotic Nitrogen Fixation in Plants (P. S. Nutman, ed.), Cambridge Univ. Press, London, pp. 319–333 (1976).

J. D. Mahon, Plant Physiol., 60:817–821 (1977).

J. D. Mahon and S. O. Salminen, Plant Soil, 56:335–340 (1980).

C. Marques Pinto, P. Y. Yao, and J. M. Vincent, Aust. J. Agric. Res., 25:317–329 (1974).

L. A. Materon and J. M. Vincent, Field Crops Res., 3:215–224 (1980).

J. McFerson, F. A. Bliss, and J. C. Rosas, in: Biological Nitrogen Fixation Technology for Tropical Agriculture (P. H. Graham et al., ed.), CIAT, Cali, Columbia, in press.

F. R. Minchin and J. S. Pate, J. Exp. Bot., 24:259–271 (1974).

F. R. Minchin, R. J. Summerfield, and A. R. J. Eaglesham, Trop. Agric. (Trinidad), 55:107–115 (1978).

F. R. Minchin, R. J. Summerfield, P. Hadley, E. G. Roberts, and S. Rawsthorne, Plant Cell Environ., 4:5–26 (1981).

D. N. Munns, R. L. Fox, and B. L. Koch, Plant Soil, 45:591–601 (1977).

L. R. Mytton, Ann. Appl. Biol., 80:103-107 (1975).
L. R. Mytton, M. H. El Sherbeeny, and D. A. Lawes, Euphytica, 26:
    785-791 (1977).
P. T. C. Nambiar, P. J. Dart, S. N. Nigam, and R. W. Gibbons, in:
    Biological Nitrogen Fixation Technology for Tropical Agricul-
    ture (P. H. Graham et al., ed.), CIAT, Cali, Colombia, in
    press.
D. Nangji, Agron. J., 72:403-406 (1980).
M. C. P. Neves, in: Biological Nitrogen Fixation Technology for
    Tropical Agriculture (P. H. Graham et al., ed.), CIAT, Cali,
    Colombia, in press.
D. O. Norris, Emp. J. Exp. Agric., 27:87-97 (1959).
D. O. Norris and L. t'Mannetje, E. Afr. Agric. For. J., 29:214-235
    (1964).
P. S. Nutman, Heredity, 3:263-292 (1949).
P. S. Nutman, Heredity, 8:35-46 (1954a).
P. S. Nutman, Heredity, 8:47-60 (1954b).
P. S. Nutman, Biol. Rev. Camb. Phil. Soc., 31:109-151 (1956).
P. S. Nutman, Heredity, 11:157-173 (1957).
P. S. Nutman, Proc. Roy. Soc. London Ser. B, 156:122-137 (1962).
P. S. Nutman, Aust. J. Agric. Res., 18:381-425 (1967).
T. Oinuma, Seitbusu, 3:155 (Cited by Torrey and Barrios, Carylogia,
    22:47) (1948).
J. S. Pate, Annu. Rev. Plant Physiol., 3:313-340 (1980).
V. R. Rao, N. S. Subba Rao, and K. G. Mukerji, Plant Soil, 39:449-
    452 (1973).
R. J. Rennie and G. A. Kemp, Euphytica, in press.
A. C. Robinson, J. Aust. Inst. Agric. Sci., 33:207-209 (1967).
P. P. Robinson, R. A. Date, and R. G. Megarrity, Aust. J. Agric.
    Res., 27:381-389 (1976).
R. J. Roughley, P. J. Dart, P. S. Nutman, and P. A. Clarke, J. Exp.
    Bot., 21:186-194 (1970).
A. D. Rovira, Bot. Rev., 35:35-57 (1969).
O. P. Rupela and P. J. Dart, in: Biological Nitrogen Fixation Tech-
    nology for Tropical Agriculture (P. H. Graham et al., ed.),
    CIAT, Cali, Colombia, in press.
M. W. Seetin and D. K. Barnes, Crop Sci., 17:783-787 (1977).
D. Sen and R. W. Weaver, Plant Sci. Lett., 18:315-318 (1980).
S. M. Souto, A. C. Coser, and J. Dobereiner, Proc. V. Reuniao
    Latinoamericano de Rhizobium, Rio de Janiero, pp. 78-91 (1970).
J. M. Spain, in: Plant Adaptation to Mineral Stress in Problem
    Soils (M. J. Wright, ed.), Agency for International Develop-
    ment, Washington, D.C., pp. 213-222 (1976).
A. Tanaka and K. Fujita, J. Fac. Agric. Hokkaido Univ., 59:145-238
    (1979).
J. A. Thompson, Nature (London), 187:619-620 (1960).
A. W. S. M. Van Egerat, Meded. Landbouwhogeschool Wageningen 72-27,
    90 pp (1972).
G. Vest, Crop Sci., 10:34-35 (1970).

G. Vest and B. E. Caldwell, Crop Sci., 12:692–693 (1972).
D. R. Viands, C. P. Vance, G. H. Heichel, and D. K. Barnes, Crop
     Sci., 19:905–908 (1979).
J. M. Vincent, Aust. J. Agric. Res., 5:55–60 (1974).
J. M. Vincent, in: The Biology of Nitrogen Fixation (A. Quispel,
     ed.), North Holland Publishing Co., Amsterdam pp. 265–341
     (1974).
F. A. Wahab, C. K. John, W. J. Broughton, Seed Technology in the
     Tropics, pp. 91–96 (1976).
L. Waters, P. J. Breen, H. J. Mack, and P. H. Graham, J. Am. Soc.
     Hort. Sci., 105:424–427 (1980).
G. Whiteaker, G. G. Gerloff, W. H. Gabelman, and D. Lindgren, J.
     Am. Soc. Hort. Sci., 101:472–475 (1976).
L. F. Williams and D. F. Lynch, Agron. J., 46:28–29 (1954).
R. J. Wilson, J. W. Burton, J. A. Buck, and C. A. Brim, Plant
     Physiol., 61:838–841 (1978).
L. Wipf and D. C. Cooper, Proc. Nat. Acad. Sci. (U.S.), 24:87–91
     (1938).
J. C. Wynne, S. T. Ball, G. H. Elkan, T. G. Isleib, and T. J.
     Schneeweis, in: Biological Nitrogen Fixation Technology for
     Tropical Agriculture (P. H. Graham et al., ed.), CIAT, Cali,
     Colombia, in press.
R. S. Yost and R. L. Fox, Agron. J., 71:903–908 (1979).
K. W. Zary, J. C. Miller, R. W. Weaver, and L. W. Barnes, J. Am.
     Soc. Hort. Sci., 103:806–808 (1978).

SPECIFICITY IN THE LEGUME-Rhizobium Symbiosis

Ivan A. Casas

Departamento de Biologia
Facultad de Ciencias Veterinarias
Universidad del Zulia
Venezuela

Host specificity has been taken as the main characteristic for the classification of species within the genus Rhizobium. Plant afinity, however, does not indicate genetic homology among Rhizobium strains. The failures in reciprocal nodulation within cross-inoculation groups, as well as cross-infection between cross-inoculation groups, substantiates the foregoing statement.

For the last eight years, studies of Rhizobium-host specificity have focused on the interactions involved in host recognition. Specific interactions between components of the macrosymbiont with those of microsymbiont have been observed and, in some cases, studied in detail. The evidence suggests that lectins (carbohydrate-binding proteins) present on legume roots and exopolysaccharides and/or lipopolysaccharides on the surface of Rhizobium may be the molecules responsible for recognition. This paper concentrates primarily on the findings regarding host-symbiont recognition for the clover-R. trifoli and soybean-R. japonicum systems.

CLASSIFICATION OF THE GENUS Rhizobium BY CROSS INOCULATION

The present classification of Rhizobium was outlined in the decades of the 1920's and 1930's. According to Bergey's Manual of Determinative Bacteriology (Buchanan and Gibbons, 1974), six species of Rhizobium are recognized. These can be distinguished by means of biochemical tests and, supposedly, by differences in infectiveness. The specificity, which is assumed for each Rhizobium species, of the

TABLE 1. Cross-Inoculation Groups (after Buchanan and Gibbons, 1974)

Group	Rhizobium species	Hosts
I	R. leguminosarum	Pisum, Vicia, Lens
I	R. trifolii	Trifolium
I	R. phaseoli	Phaseolus vulgaris, P. angustifolius, P. multiflorus
I	R. meliloti	Medicago, Melilotus, Trigonella
II	R. lupini	Lupinus, Ornithopus
II	R. japonicum	Glycine
II	Unclassified	Cowpea group

bacteria to nodulate only with particular legumes led to the formation of the cross-inoculation groups (Table 1). Each species of Rhizobium, according to this classification, nodulates only plants within a given cross-inoculation group. Conversely, within such a group, a Rhizobium from any given plant should nodulate all other plants (e.g., R. meliloti should nodulate only plants from the genera Medicago, Melilotus, and Trigonella) and a particular strain isolated, let us say, from Medicago sativa should nodulate all species of the genera Medicago, Melilotus, and Trigonella.

The criteria used for the classification based on cross-inoculation groups have been continuously challanged since the earliest 1930's. Demonstrations of failures in reciprocal nodulation as well as promiscuity among the groups have seriously questioned the validity of the classification. Consequently, alternate classifications have been proposed (de Ley, 1968; Graham, 1975; t'Mannetje, 1967). Five species were included in the genus Rhizobium, as suggested by de Ley (1968) and supported by Graham (1975): R. leguminosarum, which includes R. phaseoli, R. trifolii, and R. leguminosarum; R. meliloti; R. japonicum, including R. japonicum, R. lupini, and the cowpea miscellany; R. rhizogenes, and R. radiobacter. The last two proposed species presently fall within the genus Agrobacterium; R. rhizogenes is A. rhizogenes and R. radiobacter includes A. tumefaciens, A. radiobacter, and A. rubi. The proposed classification contemplates a less rigid criterion for cross inoculation, and it also includes DNA base ratios and DNA-hybridization data.

Nevertheless, a serious pitfall of a classification based on cross-inoculation groups, or of the proposed modification, is the fact that only the ability to nodulate is taken as the criterion to establish the infective properties of the rhizobia, and effectiveness of the symbiosis is left out. The grouping of Rhizobium-legume associations on the basis of effectiveness in N₂ fixation has obvious advantages for practical purposes. Such a classification, however, may result in the formation of subgroups within cross-inoculation groups. A brief description of such degrees of specificity follows.

(a) R. trifolii. Strain differences in N₂ fixation have been shown among varieties of African Trifolium spp. (Norris and t'Mannetje, 1964) and Trifolium subterraneum (Gibson and Brockwell, 1968), of cultivars of T. subterraneum (Gibson, 1968), as well as for white clover (Bonish, 1980).

(b) R. meliloti. Even though the medic group is the only one that still satisfies the nodulation criteria, some degree of promiscuity has been demonstrated; e.g., with Rhizobium spp. isolated from Leucaena (Trinick, 1965, 1968). Significant cross inoculation among the host genera has been observed; however, specific host patterns for effectiveness have been suggested. Eight groups of R. meliloti strains have been shown to nodulate effectively with some host species but not with others; some hosts, like M. laciniata and Trigonella suavisima, are highly specific in that they effectively nodulate only with strains isolated from the plant host (Brockwell and Hely, 1966). Cowpea, soybean, and lupin cross-inoculation groups are those with a high rate of cross infection (Lange, 1966).

(c) R. lupini. This species includes both fast- and slow-growing strains. Specificity in effectiveness shows a great deal of variation within a genus and among host genera.

(d) R. japonicum. Strains of this microsymbiont show a high degree of specificity between genotypes of Glycine max for nitrogen content, dry weight, and grain yield (Caldwell and Vest, 1968; 1970). Effective nodulation of soybean with strains of Rhizobium is quite common (Su et al., 1980).

(e) Rhizobium spp. The so-called cowpea group includes numerous host genera that nodulate with slow-growing rhizobia. Although a good deal of cross inoculation has been observed within the group, this is not absolutely true. Studies on the effectiveness in nodulation have led to the suggestion of the following grouping: promiscuous effective (PE), less promiscuous/frequently ineffective (PI), and specific (S) (Date, 1973; Graham and Hubbell, 1975). Group PE includes legumes which nodulate effectively with Rhizobium isolated from different host genera. Members of group PI nodulate with many Rhizobium strains but often ineffectively. Members of group S show specific responses in nodulation and effectiveness, but, in some instances, subgroups may be necessary to define the responses.

(f) R. phaseoli and R. leguminosarum nodulate freely with their group-specific hosts. However, some strains of R. leguminosarum nodulate Trifolium species, and some strains of R. trifolii nodulate Pisum species (Hepper and Lee, 1979).

SPECIFICITY IN MACRO-MICROSYMBIONT RECOGNITION

The specificity shown in the legume-Rhizobium implies a physical association between the microsymbiont and the root surface of the host. A high degree of specificity in the recognition must exist in order to result in an infective process and, consequently, in nodule development. A role in the recognition and binding of the symbionts has been attributed to lectins by Bohlool and Schmidt (1974) when they suggested the lectin-recognition hypothesis. Since then, a good deal of data, sometimes conflicting, has been published. Some authors have shown a correlation between infectivity and lectin-Rhizobium binding (Bhuvaneswari and Bauer, 1978; Bhuvaneswari et al., 1977; Bohlool and Schmidt, 1974; Dazzo and Brill, 1977; Dazzo and Hubbell, 1975; Dazzo et al., 1976; Dazzo et al., 1978; Kamberger, 1979; Planque and Kijne, 1977; Napoli and Albersheim, 1980b; Sanders et al., 1978; Wolpert and Albersheim, 1976; Wong, 1980), whereas others have reported nonspecific binding (Chen and Phillips, 1976; Dazzo and Hubbell, 1975a; Kamberger, 1979; Law and Strijdom, 1977; Pueppke et al., 1980; Schultz and Pueppke, 1978; Stahlhut et al., 1981; Su et al., 1980; 1981; Wong, 1980).

It was suggested that the microbial structure responsible for recognition is either the acidic exopolysaccharide (EPS) found in the capsular and/or slime material of the rhizobia (Bal et al., 1978; Bhagwat and Thomas, 1980; Bhuvaneswari et al., 1977; Bishop et al., 1977; Bohlool and Schmidt, 1974; Bohlool and Schmidt, 1976; Calvert et al., 1978; Dazzo and Brill, 1977; Dazzo and Brill, 1979; Dazzo and Hubbell, 1975b; Dazzo et al., 1976; Dazzo et al., 1979; Kamberger, 1979; Mort and Bauer, 1980; Napoli and Albersheim, 1980b; Robertsen et al., 1981; Sanders et al., 1978; Shantaharam et al., 1980; Stacey et al., 1980; Tsien and Schmidt, 1977, 1980, 1981; Zavenhuizen et al., 1980) or a lipopolysaccharide (LPS), the O antigen (Albersheim et al., 1977; Kamberger, 1979; Kato et al., 1979; Kijne et al., 1980; Maier and Brill, 1978; Maier et al., 1978; Wolpert and Albersheim, 1976). In some cases, the action of both the EPS and LPS has been suggested, the EPS being the major binding site and the LPS a minor or secondary site (Dazzo and Brill, 1979; Bhagwat and Thomas, 1980).

Interactions of soybean lectin (SBL) with R. japonicum and clover lectin with R. trifolii have received a great deal of atten-tion from various laboratories. Anomalous results in regard to the lectin-recognition hypothesis, such as the absence of SBL in nodulating soybean varieties (Pull et al., 1978; Stahlhut and Hymowitz, 1980), adds to the complexity of understanding of the recognition phenomenon.

The specific binding of clover lectin with carbohydrate receptors of the cell surface of R. trifolii is, probably, the Rhizobium-legume system with enough experimental support to fit the lectin recognition hypotheses suggested by Bohlool and Schmidt (1974). The description of the model postulated by Dazzo and Hubbell (1975b) includes the participation of clover lectin ("trifoliin," Dazzo et al., 1978), a capsular polysaccharide receptor of R. trifolii (Dazzo and Brill, 1979; Dazzo and Hubbell, 1975b), and a cross-reactive antigen (CRA) present in both infective R. trifolii capsular acidic exopolysaccharide and clover root cell surface (Dazzo and Hubbell, 1975b). R. trifolii-specific CRA was found to be present only on clover root surface (Dazzo and Brill, 1979). The clover lectin (trifoliin) that recognizes these surfaces antigens (CRA) cross-bridges them to form a correct interfacial structure that allows for specific adhesion of the bacteria to the root surface (Dazzo and Hubbell, 1975b).

Trifoliin, the Clover Lectin

An agglutinating nondialyzable factor present in preparations of clover seed lectin was reported (Dazzo and Hubbell, 1975b), purified, partially characterized, and named "trifoliin" (Dazzo et al., 1978). Inactivation of this factor by heat, pronase, trypsin, periodate, HCl at pH 3, NaOH at pH 12, and 7 M urea indicates that it is a protein (Dazzo and Hubbell, 1975b). Additional proof includes absorbance at 280 nm, staining with Comassie blue R-250, and data on immunodiffusion of purified trifoliin (Dazzo et al., 1978). Molecular-weight estimation in SDS gels gives a value of 50,000 for trifoliin. The presence of sugars in its structure was shown by staining the single sharp band in polyacrylamide gel electrophoresis with periodate-Schiff's reagent and by quantitative determinations, which yield 6.54 ± 2.56 μmol reducing sugar per mg of protein (Dazzo et al., 1978). Such data clearly indicate that trifoliin is a glycoprotein, as hinted at by Dazzo and Hubbell (1975b).

The presence of a multivalent lectin (the agglutinin present in seed extracts of clover) on the surface of clover roots was proposed (Dazzo and Hubbell, 1975; Dazzo et al., 1976), and it was later demonstrated to be trifoliin, the clover seed lectin (Dazzo et al., 1978). Trifoliin molecules are not covalently bound to seedling roots, since they are eluted from the root surface by 30 nM 2-deoxyglucose (Dazzo et al., 1976; Dazzo and Brill, 1977). The agglutinin obtained in this way and isolated by preparative gel electrophoresis was identified as trifoliin by the following criteria: identical electrophoretical mobility in native and SDS-polyacrylamide gels compared with trifoliin isolated from seeds,

immunoprecipitation with anti-trifoliin, agglutination, and binding
to R. trifolii which was inhibited specifically by 2-deoxyglucose
(Dazzo and Brill, 1977).

The presence of compounds antigenically related to trifoliin
was studied by quantitative immunofluorescence of seedling roots incubated with anti-trifoliin antiserum. Roots of Trifolium repens
var. Lousiana Nolen (the source of trifoliin for his evaluation),
T. repens var. WC-3-5, T. pratense var. C-116, and T. fragiferum
var. Salina, which are members of the same inoculation group
(Norris and t'Mannetje, 1964), did bind anti-trifoliin antibody.
Roots of T. semipilosum var. 234412, which belongs to a different
clover cross-inoculation group (Norris and t'Mannetje, 1964), bound
only low levels of anti-trifoliin antibody. The immunofluorescence
found in roots of Medicago sativa var. Florida 66 and Vernal,
Aeschynomene americana, and Lotus corniculatus var. Empire BT-15
was nil and compared only with the preimmune serum control. These
findings show that trifoliin or antigenically related compounds are
present only on the root surface of clovers but not in the other
legumes studied (Dazzo et al., 1978).

The distribution of trifoliin on roots of clover seedlings was
detected by indirect immunofluorescence, a large concentration of
the clover lectin being found in the root hair region. A closer observation revealed a gradient of fluorescence along the root hair
surface, with a maximum intensity at the root hair (Dazzo et al.,
1978). However, the distribution of the CRA was uniform on root
hairs as well as on undifferentiated epidermal root cells (Dazzo
and Brill, 1979; Dazzo and Hubbell, 1975b), and this CRA probably
was reponsible for the massive adsorption of trifoliin-coated R.
trifolii (Dazzo et al., 1976).

The specific agglutinating properties of trifoliin were known
even before its purification. All 9 strains of R. trifolii (infecting T. repens) tested were agglutinated by trifoliin, whereas the
4 noninfective R. trifolii strains and the heterologous Rhizobium
strains did not (Dazzo and Hubbell, 1975b). Similar agglutination
patterns were obtained by Dazzo and Brill (1977) with root-eluted
trifoliin, and by Dazzo et al. (1978). When the latter authors used
purified trifoliin in conjunction with 34 Rhizobium strains, only
strains infective to T. repens were agglutinated by trifoliin. Purified trifoliin agglutinates R. trifolii at concentrations as low as
0.1 to 0.2 µg protein per ml. Attachment of trifoliin to bacterial
cells was studied by means of indirect immunofluorescence. Binding
of trifoliin was confirmed in R. trifolii, whether infective or noninfective to T. repens, but not in the other Rhizobium species tested.
This specific binding was prevented by 2-deoxyglucose (Dazzo and
Hubbell, 1975b).

Agglutination of R. trifolii by crude trifoliin preparations

is inhibited by 30 nM concentrations of either 2-deoxyglucose or
N-acetylgalactosamine but not by α-D-glucose, β-D-glucose, α-D-
galactose, α-D-glucuronic acid, α-D-galacturonic acid, cellobiose,
D-mannose, D-fructose, D-xylose, α-methyl-D-mannoside, L-sorbose,
or N-acetylglucosamine (Dazzo and Hubbell, 1975b). 2-Deoxyglucose
also inhibits the agglutination of R. trifolii by purified trifoliin
(Dazzo et al., 1978). The same hapten elicits the elution of trifoliin from clover seedling roots (Dazzo et al., 1976, 1978).

Nature of Trifoliin Receptor of R. trifolii

The interaction of encapsulated R. trifolii with crude and purified trifoliin preparations has been demonstrated, and the role of
the acidic polysaccharide component of the capsular material as the
lectin receptor has been suggested (Bishop et al., 1977; Dazzo and
Brill, 1977, 1979; Dazzo and Hubbell, 1975b; Dazzo et al., 1978,
1979). The presence of surface antigens that cross-react with clover
roots (trifoliin) was shown by Dazzo and Hubbell (1975b) in both infective and noninfective strains of R. trifolii by agglutination and
indirect immunofluorescence techniques. Infective strains, however,
demonstrated (in both agglutination and indirect immunofluorescence)
a greater degree of antigenic reactivity than noninfective strains.

Actual attachment of R. trifolii to clover root hairs was shown
by electron micrography. The physical association of strain NA-30,
by means of its fibrillar capsule, with electron dense globular aggregates lying on the external surface of root-hair cell walls was
clearly depicted by Dazzo and Hubbell (1975b). Binding of FITC-
conjugated capsular material from R. trifolii strains 0403 and T37
to clover roots hairs was observed by fluorescence microscopy. The
extent of fluorescence was great at the root hair tips and decreased
toward the root hair base. Little or no binding of capsular material to other epidermal cells of clover roots was observed. FITC-
conjugated capsular material from R. meliloti F28 bound alfalfa root
in a similar way, but not clover root hairs; conversely, the FITC-
conjugated polysaccharide of R. trifolii did not bind alfalfa root
hairs (Dazzo and Brill, 1977). Competition experiments demonstrated
that both the polysaccharides and their FITC-derivatives bind to the
same receptor sites on the root hairs (Dazzo and Brill, 1977).

That the CRA of R. trifolii is associated with the capsular
material was shown by Dazzo and Hubbell (1975b). Acidic polysaccharides isolated from R. trifolii strain 403 cross-reacted with
anti-white clover root antiserum as demonstrated by immunofluorescence and immunoprecipitation in capillary tubes. The antigenic
cross-reactivity of strain 403 capsules was eliminated by treatment
with either HCl (pH 3), NaOH (pH 12), periodate, or lysozyme. After
removal of the capsular material from R. trifolii strain 403 with
periodate treatment, the underlying cell surface was still cross-

reactive with clover roots (Dazzo and Hubbell, 1975b). Binding of crude trifoliin eluted from clover roots to capsular material of R. trifolii strain 0403 was observed by passive hemagglutination using rabbit erythrocytes coated with capsular material isolated from R. trifolii (Dazzo and Brill, 1977). The possible participation of R. trifolii flagella in host recognition was ruled out (Napoli and Albersheim, 1980a).

The isolation and partial characterization of the capsular material of R. trifolii strains 403 and Bart A (which are infective and noninfective to T. repens, respectively) indicated that the capsular materials of these strains are quite similar. They show an amorphous noncrystalline structure of high molecular weight (greater than 4.6×10^6 dalton) and are devoid of cellulose microfibrils. They share properties of acidic polysaccharides and have a net negative charge at pH 8.6, and the infrared spectra of both are identical and are typical of a β-glycoside carbohydrate structure rich in hydroxyl and carboxyl groups. Nucleic acids and proteins were not detected, but trace amounts of lysine, aspartic acid, threonine, serine, glutamic acid, glycine, alanine, and isoleucine were detected in acid hydrolyzates of strain 403 (Dazzo and Hubbell, 1975b). The major component of the capsular material is carbohydrate and contains 68% neutral sugar and 32% uronic acid. The sugar components determined by gas–liquid chromatography are 2–deoxyglucose, α-D-galactose, α- and β-D-glucose, α-D-glucuronic acid, and one unidentified compound (Dazzo and Hubbell, 1975b). The presence of those sugars was confirmed by paper chromatography (Dazzo and Brill, 1979). Isolation and separation by ion–exchange and gel filtration chromatography of capsules of infective R. trifolii strains 0403 and 2S-2 yield three carbohydrate-containing fractions: A, B, and C. Strains 0403 has abundant capsular material, and strain 2S-2 is a nonmucoid mutant containing a granular "microcapsule." Only fractions A and B (a) show a positive capillary precipitin test with anti-clover root antibody, which was confirmed with quantitative analysis of antibody protein in immunoprecipitates, (b) have the ability to neutralize the agglutinating activity of anti-clover root antiserum, and (c) inhibit (equally and exclusively) the passive hemagglutination of rabbit erythrocytes by anti-clover root antiserum. Fraction C from capsular material of R. trifolii and all three fractions separated from the same cellular component of R. melliloti strain 102F28 do not display the same immunoreactions of capsular fractions A and B from R. trifolii. Ring precipitation was observed in capillary tubes containing trifoliin and fraction A or B of the capsular polysaccharides from R. trifolii strain 0403. Again fraction C from this rhizobium and all three capsular fractions from R. meliloti strain 102F28 did not precipitate with trifoliin (Dazzo and Brill, 1979).

Capsular polysaccharides fractions A and B but not fraction C from R. trifolii strains 0403 or 2S-2 show competitive binding with unfractioned FITC-capsular polysaccharides (Dazzo and Brill, 1979), which specifically binds to clover root hairs (Dazzo et al., 1978). Antibodies against fractions A or B, when allowed to react with clover roots, exhibit a uniform distribution on the surface of both differentiated and undifferentiated epidermal cells of the root hair region when assayed by indirect immunofluorescence (Dazzo and Brill, 1979). These results clearly indicate that fractions A and B contain the CRA which specifically binds to the trifoliin molecules, which in turn are attached to the CRA on the root hairs, and the CRA is uniformly distributed on both undifferentiated epidermal root cells and root hairs (Dazzo and Brill, 1979).

Immunoelectrophoresis of purified capsular material of R. trifolii 0403 and its fractions A, B, and C with anti R. trifolii 0403 (sonically oscillated) cells shows clear patterns of identity between fraction A and the fast migrating antigen of the capsular material and between fraction B and the slow migrating antigen. Fraction C does not show any preciptin band with the test antiserum (Dazzo and Brill, 1979). Tests of purity of each of the purified polysaccharide fractions A, B, and C of R. trifolii 0403 indicate the absence of protein, DNA, RNA, glycoprotein, and phosphoprotein. Acid hydrolyzates of fractions A and B yielded identical chromatograms of 9 silver nitrate-reactive compounds. Four of these spots correspond to those of glucuronic acid, galactose, glucose, and 2-deoxyglucose standards. Fraction C showed only four spots, three of which were present in acid hydrolyzates of fractions A and B. LPS components, endotoxic lipid A, 2-keto-3-deoxyoctanoate (KDO), and heptose were found to be present in fraction B and absent from either fractions A or C (Dazzo and Brill, 1979).

Transfer of genetic material responsible for the synthesis of capsular material of R. trifolii has been accomplished using DNA of R. trifolii strain 0403 and Nif strains of Azotobacter vinelandii as recipients in transformation experiments (Bishop et al., 1977; Dazzo and Brill, 1979). Nif$^-$ strains UW1, UW6, and UW10 of A. vinelandii were transformed to Nif$^+$ with DNA isolated from R. trifolii strain 0403 or R. japonicum strain 61A76. The frequencies of reversion (from 1.1×10^{-8} to 6.3×10^{-8}) and transformation (from 1.5×10^{-6} to 2.4×10^{-6}), cultural and morphological characteristics, and specific phage sensitivity demonstrated that the Nif$^-$ isolates were indeed A. vinelandii transformants. The same 6 strains out of 46 Nif$^+$ transformants tested were agglutinated by both trifoliin and antiserum against clover root antigen. Trifoliin and antiserum against clover root antigen did not agglutinate the remaining 40 Nif$^+$ transformants of the 10 Nif$^+$ A. vinelandii transformants obtained with R. japonicum DNA. Indirect immunofluorescence confirmed the specific binding of antiserum against clover root antigen to the cell surface of A. vinelandii transformant strains Rt Av10-54 and RtAv10-29 (Bishop et al., 1977). Attachment of A. vinelandii hybrid transformant RtAv10-54 to clover root hair tips was shown by microscopic observation, and the adherence was prevented by the addition of 30 mM of 2-deoxyglucose to the cell suspension (Dazzo and Brill, 1979).

The binding of R. trifolii strain 0403 to trifoliin and its adherence to clover root epithelial surfaces change with cell culture age (Dazzo et al., 1979). Agglutination of R. trifolii strain 0403 cells with trifoliin changed with the age of agar cultures, as did the binding of trifoliin as well as the attachment of rhizobium cells to clover root hairs (Dazzo et al., 1979). The highest specific agglutination activity of trifoliin was found with cells grown for 5 days on defined agar medium; trifoliin-agglutinating activity was roughtly <5%, 10%, 50%, and <5% of the maximum, obtained at 5 days of growth, for cells cultured for 3, 4, 6, and 7 days, respectively. Transmission electron microscopy of cells stained with ruthenium red revealed that 3- and 7-day-old cultures had little or no capsular material, whereas abundant capsular material was present in cells of 5-day-old cultures (Dazzo et al., 1979). The binding of trifoliin to cells from various culture ages was examined by indirect immunofluorescence. Binding of trifoliin was associated with the presence of capsular material: 3-day-old cells were devoid of capsules and hence no cell-bound trifoliin was observed; 5-day-old cells, which had plenty of capsular material, showed a uniform trifoliin adherence of their surface; 7-day-old cells were unencapsulated and did not bind trifoliin. Direct microscopic observations showed cells from 5-day-old cultures attached to clover root hairs in larger numbers than cells from 3- to 7-day-old cultures (Dazzo et al., 1979).

Studies of the effect of age of R. trifolii in liquid culture on the appearance of lectin receptors and the attachment to clover root hairs revealed a pattern similar to that observed with agar cultures (Dazzo et al., 1979). The presence of CRA on the cell surface of R. trifolii 0403 at various growth stages in synthetic broth was examined by means of indirect immunofluorescence. The results showed two periods during the growth cycle when attachment to anticlover root antigen was maximum (about 90% of the cells were reactive with antiserum), the first one just after cells left the lag phase and the second at the early stationary phase. Cells that were entering the stationary phase also attached to clover root hairs in higher numbers (Dazzo et al., 1979).

Further Considerations of the Clover-R. trifolii Recognition

All the evidence accumulated so far clearly points to the acidic polysaccharide component of the capsular material as the structure which contains the chemical determinants for the trifoliin receptor. The data strongly suggest that this receptor may be the CRA (or a determinant closely associated with the CRA) on the cell surface of

R. trifolii. One fraction of the trifoliin receptor-containing
acidic polysaccharide could be associated with LPS, which, under
certain conditions, replaces the neutral O antigen polysaccharide
at the core-lipid A structure (Dazzo and Brill, 1979).   Purified
O antigen-LPS, however, does not contain the 2-deoxyglucose (Carlson
et al., 1978; Zevenhuizen et al., 1980), which is considered to be
the specific hapten in the clover-R. trifolii recognition process
and for the related immunoreactions (Dazzo and Brill, 1977, 1979;
Dazzo and Hubbell, 1975b; Dazzo et al., 1976, 1978).   Carlson et al.
(1978) found no correlation between the nodulation group to which
the Rhizobium belongs and the chemical composition or immunochem-
istry of their lipopolysaccharides. Conversely, purified acidic
polysaccharides of R. trifolii strains 0403 (encapsulated) and 2S-2
("microencapsulated") have the determinants for trifoliin binding
and the anticlover root CRA, although (some) qualitative differences
of their monosaccharide components were demonstrated (Dazzo and
Brill, 1979).

     The interaction of R. trifolii with clover root hairs results
in two distinct patterns: (a) a specific adsorption of infective
R. trifolii (averaging 25.8 ± 5.9 cells per 200 μm of root hairs),
which is prevented by nM 2-deoxyglucose and (b) a nonspecific adsorption of infective and noninfective R. trifolii and R. meliloti
(2.60 ± 1.06, 5.6 ± 1.9, and 2.44 ± 1.0 to 2.90 ± 1.10 cells per
200 μm of root hairs, respectively) not affected by 2-deoxyglucose
(Dazzo et al., 1976).   Nodulation of white clover by R. trifolii
0403 and the adsorption of this infective strain to the surface of
clover roots hairs are affected by the presence of 16 nM nitrate or 1
nM ammonium, which completely inhibit infection and nodulation and
the adsorption of R. trifolii to root hairs (Dazzo and Brill, 1978).
It is important to point out that, at the levels of nitrate (16
nM) and ammonium (1 mM) that completely inhibit infection, the adsorption of cells to root hair surfaces drops to 2 to 4 from 24
cells adsorbed per 200 μm of root hair in the absence of fixed nitrogen (Dazzo and Brill, 1978).   These figures correlate well with
the nonspecific and specific binding of R. trifolii to root hair
surfaces (Dazzo et al., 1976).

     Two closely related substrains of R. trifolii strain NA34, one
a noninvasive (nonnodulating) mutant (SU847), the other an infective (non-nitrogen-fixing) mutant (SU846), when mixed give rise to
effective (N₂-fixing) nodules. No transference of genetic material
between the two substrains could be demonstrated (Rolfe et al.,
1980).   The evidence indicates that the mutant SU847, which has a
defective Roa (root adhesion) (Vincent, 1978) phenotype, actually
infects the host (Rolfe et al., 1980). How this infection was
brought about is not known, but somehow this strain has been able
to overcome the defective genetic trait at the recognition stage
(root adhesion).

Relevant R. trifolii mutants, such as SU846 and SU847, as well
as relevant genetically modified bacteria, such as the A. vinelandii
transformants with modified capsular material obtained with R.
trifolii DNA (Bishop et al., 1977; Dazzo and Brill, 1979) may help
to fully understand the recognition process in the clover–R. trifolii
system.

INTERACTION OF SOYBEAN LECTIN WITH Rhizobium japonicum

Unlike the evidence that favors the lectin recognition hypothesis for the clover–R. trifolii symbiosis, data on the soybean
lectin (SBL)–R. japonicum interaction are still incomplete for a
full understanding of the host–symbiont recognition.

Using soybean lectin conjugated with fluorescent dye, Bohlool
and Schmidt (1974) studied the interaction of lectin with Rhizobium
cells. Their results show that fluorescein isothiocyanate (FITC)-
labeled SBL binds to all but 3 of the 25 R. japonicum strains used.
Their report also indicated that a marked FITC–SBL is observed about
one end of the R. japonicum cells. The observations of the spe-
cifcity of binding of seed SBL to R. japonicum cells were confirmed
by Bohlool and Schmidt (1976), Bhuvaneswari et al. (1977), and
Bhuvaneswari and Bauer (1978). Affinity chromatography with re-
purified FITC–SBL was used to demonstrate that 15 of 22 strains of
R. japonicum bind FITC–SBL. The remaining 7 strains do not nodulate
soybean. Binding controls run with 9 strains of other rhizobia did
not show detectable levels of fluorescence. FITC–SBL binding of R.
japonicum cells is inhibited by N-acetyl-D-galactosamine at concentrations greater than 1 mM. Only D-galactose, out of 11 sugars
tested, was an effective hapten of FITC–SBL binding to R. japonicum
strain 3I16-138 cells (Bhuvaneswari et al., 1977).

A study of the influence of root exudate and rhizoplane culture conditions on the development of SBL receptors by various R.
japonicum strains was carried out by Bhuvaneswari and Bauer (1978).
All 6 of 11 R. japonicum strains that fail to bind FITC–SBL when
grown in synthetic medium developed specific receptors for the
lectin when cultured in association with roots of soybean seedlings.
Five of those 6 strains also bound FITC–SBL when cultured in soybean root exudate. However, the appearance of SBL receptors was
evident in 2 Rhizobium strains which did not nodulate with soybean.
Other heterologous rhizobia, when grown in either of the described
media, did not show any hapten-reversible binding of FITC–SBL. On
the other hand, root exudate of pea seedlings sustained the development of specific SBL receptors by 3 strains of R. japonicum; two of
those strains, 61A76 and 61A72, did not bind SBL lectin when grown
in synthetic medium, but the other one did. These findings suggest
that there may a rather wide range of plant species which provide
suitable environments for the elicitation of SBL receptor development (Bhuvaneswari and Bauer, 1978).

Additional information on the lack of correlation between nodulation and binding of the homologous host lectin was shown by Chen
and Phillips (1976) and by Law and Strijdom (1977). The binding of
R. japonicum and other Rhizobium species to pea roots and to FITC-
pea lectin as well as the binding of FITC-SBL to R. leguminosarum
128C53, R. meliloti 102F51, R. trifolii 162x68 (and to a lesser extent to R. phaseoli 127K17 and R. spp 22A1 but not to R. japonicum
61A96) was reported by Chen and Phillips (1976). Law and Strijdom
(1977) added the following information: with the exceptin of 2 of
10 R. japonicum strains and 6 of 10 R. phaseoli strains, none of the
remaining 48 Rhizobium strains (10 of R. leguminosarum, 5 of R.
trifolii, 5 of R. meliloti, 5 of R. lupini, 3 miscellaneous rhizobia,
and 10 specific Rhizobium strains for Asphalathus linearis and 10
for Lotononis bainesii) bound lectin from their normal nodulating
host. Heterologous host lectin-Rhizobium attachment was observed
(e.g., A. linearis and P. vulgaris lectins to R. japonicum strain
WV61).

The information given thus far indicates the complexity of the
lectin-mediated host-symbiont recognition in soybean. A more detailed account of properties of SBL and its receptor(s) on the surface of R. japonicum will follow.

Presence of Lectins in Seeds and Other Tissues
of Soybeans

The lack of seed lectins in 5 of 102 lines of Glycine max (L)
Merr, which nevertheless nodulated with R. japonicum strains 3I1b138,
505 W, and 61A72, was reported by Pull et al. (1978) and Su et al.
(1980). These authors also corroborated the presence of the 92,000
dalton isolectin in seeds of soybean line D68-127 (Fountain and Yang,
1977). Su et al. (1980), however, found evidence suggesting that
line D68-127 also contains the 120,000 dalton lectin, and the earlier
report of the 92,000 dalton lectin may be a result of either protein
decomposition during purification or of an unreliable method for molecular-weight determination. Isolectins in soybean have also been
shown by other authors (Catsimpoolas and Meyer, 1969; Lis et al.,
1966).

Evidence for the absence of the 120,000 dalton SBL in soybean
lines Columbia, Norredo, Sooty, T102, and Wilson-5 includes polyacrylamide gel electrophoresis, hemagglutination, FITC-SBL binding
to R. japonicum (Pull et al., 1978), immunodiffusion, and radioimmunoassay (Su et al., 1980). The percentage recovery of added
tritium-labelled SBL from seed meal of the five lines rules out the
possibility of a failure of lectin detection resulting from the inactivation of the lectin during extraction (Pull et al., 1978).

In a screening of 2624 accessions of the U.S. Dept. of Agriculture soybean germplasm collection, Stahlhut and Hymowitz (1980)

identified 13 accessions lacking SBL. The absence of seed lectins
is apparently genetic. Orf et al. (1978) demonstrated that the
presence of soybean lectin in line T-102 is controlled by a single
dominant gene, designated Le. The homozygous recessive "le le"
results in the absence of SBL.

Seed lectin in Glycine soja, considered on cytogenetic evidence
to be the wild ancestor of Glycine max (Hymowitz, 1980), was studied
by Stacey et al. (1980) and Stahlhut et al. (1981). The seed lectins of G. soja and G. max exhibit antigenic homology as determined
by the Ouchterlony double-diffusion assay (Sacey et al., 1980;
Stahlhut et al., 1981) as well as by having the same electrophoretic
mobility in acrylamide gel electrophoresis (Stahlhut et al., 1981)
and by identical peptide maps made by limited proteolysis and gel
electrophoresis analysis (Stacey et al., 1980). All these properties strongly suggest that the lectins of the two species have a
very similar, if not identical, structure. The binding of G. soja
lectin with R. japonicum is inhibited and reversed by the same
haptens (D-galactose and N-acetyl D-galactosamine) of the SBL
(Stacey et al., 1980).

The absence of G. soja seed lectin was reported for 49% of 559
accessions of the U.S. Dept. of Agriculture G. soja collection by
the double-diffusion assay and by polyacrylamide gel electrophoresis.
However, the lectinless accessions nodulated equally well as those
that did contain lectins (Stahlhut et al., 1981).

The presence or absence of seed lectins does not necessarily
imply that the same protein may be present in the roots, where the
actual host-symbiont recognition takes place. There is some information in regard to the presence of lectins in various soybean
tissues.

To quantify the levels of SBL, Pueppke et al. (1978) used
three different analytical approaches: hemagglutination assay,
radioimmunoassay (^{125}I-SBL) and isotope dilution (^{3}H-SBL). Soybean lectins were found to be present in embryo axes, cotyledons,
and seed coats of soybean seeds, the cotyledons containing the
larger percentage of lectin. The levels of SBL in young plants decreased with time. At abscission (16 to 18 days) the quantity of
SBL fell to values between 0.1 and 1.0 µg/g fresh weight from levels
above 5000 µg SBL/g fresh weight found at planting. When primary
roots, secondary roots, and leaves were examined separately, readily
detectable levels were found in 4-day-old seedlings. The detectable
levels of SBL decreased until 15 to 16 days after planting, when it
became nil. The limit of detection for the assay was about 0.1 µg/g
fresh weight of tissue (Pueppke et al., 1978).

In this regard, rather conflicting results were presented by
Su et al. (1980). Time-course experiments from 3 to 30 days with

growing plants of lines Horosoy 63 and Norredo and analysis of 5-
day-old roots and hypocotyls of 20 lines (including Horosoy 63 and
Norredo) failed to demonstrate the presence of SBL by immunoagglutination and radioimmunoassay in roots and hypocotyls. Four-
day-old Beeson seedlings, one of the 20 soybean lines analyzed by
Su et al. (1980), were previously reported to possess detectable
quantities of SBL in primary roots, secondary roots, and leaves
(Pueppke et al., 1978). Occasional false positive SBL determinations are avoided when stringent care is taken to preclude cross-
contamination by seed or cotyledon meals (Su et al., 1980).

    In any event, the papers of Pueppke et al. (1978) and Su et al.
(1980) dealt with phosphate-buffered saline-soluble SBL. Root hair-
bound lectin, if present, was not detected by the quantitative assays used by those authors. Stacey et al. (1980) demonstrated the
presence of SBL soja seed lectins or a protein that was immunologically cross-reactive with those lectins on the surface of soybean
variety Corosoy and Glycine soja roots. Four-day-old soybean and
G. soja roots treated with anti-SBL but not with preimmunoserum did
bind immunolatex beads prepared with goat anti-rabbit Ig G (Stacey
et al., 1980). Evidence for the presence of membrane-bound lectins
in particular fractions of soybean root homogenates has been reported. Nevertheless, some properties of this membrane-bound lectin
suggest that it is a different protein than the seed soybean lectin
(Keegstra and Andrews, 1978). The membrane-bound lectin does not
cross-react with anti-SBL and is soluble in detergent, and its hemag-
glutinating activity is inhibited by serum glycoproteins, salicin,
and p-nitrophenylglucosaminide.

    The existing evidence does not prove nor negate the involvement
of soybean lectins in host-symbiont recognition. As yet, there is
not enough information on the presence, persistence, and composition
of the lectins on the surface of soybean roots to allow for any relevant conclusions. Some data, however, demonstrate the presence of
at least two root lectins: one similar to soybean seed lectin which
binds R. japonicum in a hapten-reversible fashion (Stacey et al.,
1980), the other different from seed SBL (Keegstra and Andrews, 1978).

Nature of Lectin Receptors of R. japonicum

    Although there is not total agreement on the location and bio-
chemical nature of the lectin receptor on the surface of R. japonicum,
the existing evidence points to the acidic exopolysaccharide (EPS)
constituent of the capsular material (Bal et al., 1978; Bhuvaneswari
et al., 1977; Bohlool and Schmidt, 1974, 1976; Calvert et al., 1978;
Kamberger, 1979; Mort and Bauer, 1980; Shantaharam et al., 1980;
Stacey et al., 1980; Tsien and Schmidt, 1977, 1980, 1981; Zevenhuizen et al., 1980) rather than the O-antigen-containing lipopolysaccharides (LPS) (Albersheim et al., 1977; Kato et al., 1979; Maier
and Brill, 1978; Maier et al., 1978; Wolpert and Albersheim, 1976).

The results of interactions between the O-antigen-containing lipopolysaccharides of R. japonicum strain 505, R. leguminosarum strain 128-C-63, R. phaseoli strain 127-K-24, and Rhizobium strains 22A3 and 127-E-12 with lectins of their homologous host suggest that LPS may be the specific lectin receptors on the surface of R. japonicum. However, only a small percent of the added LPS (135% for R. japonicum, 36% for R. leguminosarum, 5 to 23% for R. phaseoli, and 25% for the others) did bind to the agarose-immobilized lectin columns (Wolpert and Albersheim, 1976). The amount of LPS which binds to the lectin column is time-dependent, and more lipopoly- saccharide binds to the column when allowed to react for 100 min rather than letting the LPS pass directly through the affinity column. Longer interaction times, however, result in a concurrent decrease in the amount of LPS bound to the column (Albersheim et al., 1977). This may be a result of enzymatic action of lectins (Alber- sheim and Wolpert, 1976) altering the LPS structure (Albersheim et al., 1977) and hence decreasing its affinity for the lectin columns. Kato et al. (1979) presented data supporting the observa- tion that R. japonicum LPS is the receptor for lectin binding. Their binding preparations contained 2-keto-3-deoxyoctonate (KDO), a compound present in purified rhizobial LPS preparations (Alber- sheim et al., 1977; Carlson et al., 1978; Mort and Bauer, 1980; Zevenhuizen et al., 1980), whereas exopolysaccharides are devoid of KDO (Dudman, 1976; Dudman, 1978; Mort and Bauer, 1980; Kamberger, 1979).

Maier and Brill (1978) associated the O antigen portion of the lipopolysaccharide with nodulation. Nonnodulating mutants had dif- ferent O antigen composition and immunoagglutination than the nodu- lating wild type. These results were corroborated by means of trans- formation experiments using Azotobacter vinelandii as the recipient cell (Maier et al., 1978). The wild-type and mutant strains used for the experiments, however, lacked extractable acidic polysac- charides on the cell surface (Maier and Brill, 1978).

Polarity of the binding of homologous fluorescent antibody and of fluorescein isothiocyanate-labeled soybean lectin (FITC-SBL) to the R. japonicum external structure was observed by Bohlool and Schmidt (1974) and Bohlool and Schmidt (1976), suggesting that polar tips of R. japonicum may contain the sites of lectin attachment. No indication of the nature of the binding sites was given at that time. The presence of extracellular polar bodies as a consequence of par- tial encapsulation during exponential growth has been described by Tsien and Schmidt (1977).

Additional evidence for polar attachment of SBL and of G. soja lectin to R. japonicum strain 3I1b110 by means of indirect immuno- latex technique has been presented (Stacey et al., 1980). R. japoni- cum cells also bind in a polar fashion to G. soja and 4-day-old soy- bean root hairs. The binding of R. japonicum to both the lectins and root hairs is inhibited by sugar haptens (N-acetyl-D-galactosamine and D-galactose) (Stacey et al., 1980).

Variation in the SBL-binding property with culture age has been observed with R. japonicum strain 3I1b-128. Maximum binding occurred in the mid logarithmic phase of growth (about 4 days); thereafter, the percentage of cells binding FITC-SBL decreases to 0 for a 9-day-old culture. During growth, the average ^{3}H-SBL molecules bound per cell increased abruptly between 24 and 36 h, then decreased abruptly by 48 h; afterward, the number of ^{3}H-SBL binding sites per cell increased to a maximum at day 4 and then again gradually decreased to 0 for cultures in the stationary phase. These observations indicate that the receptors on the cell surface may be transient and nonconstitutive. A good correlation between SBL binding and slime production was also observed (Bhuvaneswari et al., 1977).

Studies of the binding of purified ferritin-labeled soybean lectin (Fe-SBL) to cells of R. japonicum strain 3I1b-138 together with electron microscope observations of whole mounts, thin sections, and freeze-etched material showed that, of the cell surface components, Fe-SBL binds to the capsular material but not to the flagella nor the outer membrane. Capsular material is lost quite easily from cells during sample preparation for electron microscopy. The addition of ruthenium red during fixation and increasing the ionic strength of the fixation buffer improved integrity of the capsular material, though some cells appeared to be only partially encapsulated (Calvert et al., 1978). Tsien and Schmidt (1981) gave additional evidence of partial encapsulation of R. japonicum associated with binding to ferritin-SBL. The portion of encapsulated cells of R. japonicum strain 3I1b-138 increased to about 55% of the cells at the time of maximum lectin binding (1 to 3 days). Concurrent with the decrease of the percent lectin binding cells, the percentage of encapsulated cells dropped to about 30%. From day 7 to 11, the capsules became smaller and were present about one end of the cells. After 12 to 13 days, the cells were almost completely devoid of capsules (Mort and Bauer, 1980).

A biochemical and morphological explanation of transient SBL binding to R. japonicum strain 3I1b-138 has been given by Mort and Bauer (1980). They found that the chemical composition of the EPS of R. japonicum strains 3I1b-138 and 3I1b-110 was quite constant with culture age in terms of mannose, glucose, galacturonic acid, and acetate content; for galactose and 4-0-methylgalactose, the sum of the concentrations remained constant during growth. Increments of 4-0-methylgalactose concentration paralleled the decrease of D-galactose concentration. Substantial changes in the content of D-galactose and 4-0-methylgalactose in the capsules occurred at approximately the same time as the lectin-binding ability of the cells decreased and could be associated with a large drop in the percent

of encapsulated cells; thus, it is probable that a combination of
changes in capsular polysaccharide composition and percentage of encapsulated cells can explain the reported changes of lectin-binding
capacity of R. japonicum associated with culture age (Mort and
Bauer, 1980). The existence, localization, and partial characterization of soybean lectin-binding polysaccharide (LBPS) of R.
japonicum strain U.S. Dept. of Agriculture 138 have been reported
by Tsien and Schmidt (1980) and Tsien and Schmidt (1981). Hemmaglu-
tination-inhibition assay (HIA) with EPS or N-acetyl-D-galactosamine, the specific sugar hapten of SBL, give essentially a linear
dose-response. As little as 0.2 µg of EPS per ml could be detected
consistently by HIA (Tsien and Schmidt, 1980). The concentration
of LBPS in the culture increased parallel to the growth of R.
japonicum up to the end of the logarithmic phase and then increased
slowly during the stationary phase. LBPS in the supernatant fluid
also increased during the logarithmic phase and then decreased
slightly during the stationary phase. LBPS concentration in the
supernatant fluid remained high for up to 2 weeks. The HIA results
were confirmed by Ouchterlony gel-diffusion assays. The association
of LBPS with a viscous polysaccharide that is also synthetized during growth may be the cause of the decline in the hemagglutination-
inhibition activity observed in the supernatant fluid (Tsien and
Schmidt, 1980).

Subsequent studies have shown the presence of two forms of LBPS
in the EPS complex: (a) a high-molecular-weight polysaccharide component of a viscous cell-gel complex which is not diffusible in 1%
gel, and (b) a lower-molecular-weight polysaccharide dissociated
from the cells and detectable in the cell-free supernatant fluid by
Ouchterlony gel-diffusion techniques (Tsien and Schmidt, 1981).
Both LBPS forms are acidic EPS and bind SBL in a hapten-inhibiting
manner (by N-acetyl-D-galactosamine and D-galactose), and their
structural difference may be the degree of polymerization. Polar
distribution of LBPS is associated with the high-molecular-weight
form (Tsien and Schmidt, 1981). Such localization is consistent
with the polar binding of SBL reported by Bohlool and Schmidt (1974),
(1976), Calvert et al. (1978), Mort and Bauer (1980), Stacey et al.
(1980), and Tsien and Schmidt (1977). Since the capsule or partial
capsule is not tightly associated with the cell, continuous washing
and centrifugation may result in the loss of EPS capsular material
and hence the LBPS associated with it.

The purified acidic EPS, but not the purified O antigen containing LPS, or R. japonicum strain CB-1809 forms precipitin lines
specifically with SBL as determined by the Ouchterlongy double-diffusion assay. No binding of SBL to any purified LPS or EPS of R.
meliloti strains 2011 and MVII/1, R. leguminosarum LVII/1, R.
phaseoli DSM 30137, Rhizobium 32H1, R. lupini strains WU425, H13-3,
and 10319, and a pseudomonad was found. N-Acetyl-D-galactosamine
inhibited the precipitin lines. Further evidence of the interaction

of lectins with polysaccharides has been obtained by electron microscopy (Kamberger, 1979).

Remarks on Soybean-R. japonicum Recognition

     The existing evidence clearly indicates that seed SBL binds to
R. japonicum capsular EPS (Bal et al., 1978; Bhuvaneswari et al.,
1977; Bohlool and Schmidt, 1974, 1976; Calvert et al., 1978; Kamberger, 1979; Mort and Bauer, 1980; Shantaharam et al., 1980; Stacey
et al., 1980; Tsien and Schmidt, 1977, 1980, 1981; Zevenhuizen et
al., 1980). This binding is specific. Two forms of EPS, designated
LBPS, may be the polymers to which seed SBL binds (Tsien and Schmidt,
1980, 1981). Diminishing of SBL-binding capacity of R. japonicum
cells is partially a consequence of the detachment of the weakly
associated capsular material from the cell (Bohlool and Schmidt,
1974, 1976; Calvert et al., 1978; Mort and Bauer, 1980; Stacey et
al., 1980; Tsien and Schmidt, 1977, 1981). The existence of SBL
or a protein cross reacting with seed SBL on the surface of G. soja
and G. max has been demonstrated (Stacey et al., 1980). Conflicting
evidence on the existence of SBL in soybean roots has been reported
(Pueppke et al., 1978; Su et al., 1980).

     The possibility of the interaction of R. japonicum with lectins
other than the seed SBL can not be ruled out. On this matter,
Keegestra and Andrews (1978) reported the existence of a detergent-
soluble, membrane-bound lectin in soybean roots that differs from
the seed SBL. The concentration of SBL at the root hair surface may
not be large (Stacey et al., 1980); therefore, it may be difficult
to isolate and characterize the lectin. However, the use of more
suitable detection techniques, such as the immunolatex-inhibition
assay (Stacey et al., 1980), may be of help in elucidating the presence of lectins on root surfaces.

INTERACTION OF HOST LECTINS WITH Rhizobium
SPECIES OTHER THAN R. trifolii and R. japonicum

     Information on the interaction of legume lectins and Rhizobium
species other than R. trifolii is still insufficient for a reliable
model to support the possible interaction of lectins with Rhizobium
cells in the host-symbiont recognition. Nevertheless, a short description of some relevant findings on this matter will be discussed.

     Double-diffusion studies on the interaction with legume lectins
and rhizobial cell-surface antigens of 6 species of Rhizobium led
Kamberger (1979) to suggest the classification of the 6 species in
two groups. The first one has a host-specific LPS-lectin interaction (R. meliloti-Medicago sativa, R. leguminosarum-Pisum sativum/
Lens culinaris). The second group has a host-specific EPS-lectin
interaction (R. japonicum-Glycine max). The other three systems

show no specific lectin interaction with either LPS or EPS. Concavalin A formed precipitin lines with purified EPS of Rhizobium
strain 32H1, R. meliloti strains 2011 and MVH/1, R. leguminosarum
strain LHV/1, and R. phaseoli strain DSM30137 and with LPS of R.
meliloti MVH/1, R. leguminosarum LVH/1, Rhizobium 32H1, R. japonicum CB1809, and R. lupini strains 10319 and WU425. Erythroagglutinin
from Phaseolus vulgaris (kidney bean) did not precipitate with
either the LPS or the EPS from R. phaseoli. Data on R. lupini-
lupin interaction were not obtained because the lectin from lupins
was not purified (Kamberger, 1979).

     Interactions of R. leguminosarum and legume roots show that al-
though the FITC-antibody-R. leguminosarum 128C53 complex binds to
Pisum sativum root surface and root hairs, it also binds the roots
of Canavalia ensiformes, Lupinus polyphyllus, Trifolium pratense,
and Medicago sativa (Chen and Phillips, 1976). These authors also
reported that ^{32}P-labelled R. meliloti 102F51, R. trifolii 162x68,
R. phaseoli 127K17, R. japonicum 61A96, R. lupini 96B9, Rhizobium
22A1, and R. leguminosarum 128C53 did bind to pea roots even though
only R. leguminosarum could infect this plant. Data supporting the
findings of Chen and Phillips (1976) were reported by Law and Strijdom
(1977), who found no correlation between effectiveness and host-
lectin attachment to Rhizobium cells. Binding of FITC-labelled pea
lectin to 5 strains of R. leguminosarum tested was not observed.
However, different patterns of binding were found in experiments
carried out with FITC-labelled lectins of pea, lentil, broad bean,
and jack bean and 5 effective strains of R. japonicum, 2 strains of
R. phaseoli, 2 strains of R. japonicum, and 1 strain of Rhizobium
sp. (Wong, 1980). With the exception of the binding of lentil
lectin to R. japonicum strain 61A133, the lectins of lentil, pea,
and broad bean did not display heterologous binding, but only 3 of
5 R. leguminosarum strains bound to lentil lectin. Glucose, fructose, and mannose inhibited the binding of FITC-complexed lectins
to the Rhizobium cells (Wong, 1980). Specific binding of pea lectin
to R. leguminosarum was also reported by Planque and Kijne (1977).

     Specific interactions of agarose-bound host lectins and rhizo-
bial lipopolysaccharides was described for pea-R. leguminosarum,
soybean-R. japonicum, red kidney bean-R. phaseoli, and jackbean-
Rhizobium sp. (Albersheim et al., 1977; Wolpert and Albersheim,
1976). No heterologous binding was found in these experiments.
Data on rhizobial LPS structure and immunoreactions indicate that
LPS preparations are structurally diverse and suggest that there
is no obvious correlation between LPS structure and nodulation
groups. However, the data do not rule out the possibility that
structural regions in the LPS other than the immunodominant sites
could be involved in Rhizobium-legume recognition (Carlson et al.,
1978). This statement is supported by experimental evidence that
shows specific precipitin reactions between R. leguminosarum LPS
and pea lectin (Kamberger, 1979). Unfortunately the strains used

in each of these studies were not the same (Carlson et al., 1978; Kamberger, 1979). Planqué and Kijne (1977) reported that pea lectins bind to a polysaccharide in the cell wall of R. leguminosarum.

The involvement of rhizobial EPS rather than LPS in host-symbiont recognition was suggested by Sanders et al. (1978) and later supported by the data of Napoli and Albersheim (1980b). Mutant strains of R. leguminosarum 128C53 with diminished EPS exhibit a clear reduction in infection and nodulation, but no apparent change in the LPS was found (Sanders et al., 1978). Details of shepherd's crook formation, the appearance of infection treads, and nodulation indicate that those phenomena are correlated with the amount of EPS present on the cell surface (Napoli and Albersheim, 1980b). However, Kamberger (1979) found host-specific LPS-lectin interactions for the fast growing rhizobia, including R. leguminosarum, by means of an Ouchterlony double-diffusion test. Additional reports on agglutination of R. leguminosarum cells and isolated LPS by purified seed and peat root lectin (Kijne et al., 1980) favor the possible involvement of R. leguminosarum LPS as the cellular structure responsible for the host-R. leguminosarum recognition. Identical structures with respect to the glycosyl sequence and anomeric configuration of the glycosidic linkages have been shown for the acidic polysaccharides excreted by R. leguminosarum strains 128C53 and R. trifolii NA30 (Robertson et al., 1981). Cross inoculation of some strains of R. leguminosarum and R. trifolii has also been demonstrated (Hepper, 1978; Hepper and Lee, 1979).

Isolation and partial characterization of pea seed and root lectins indicate the presence of two related lectins in seeds of Pisum sativum var. Rondo seeds and the presence of one on the root surface (Kijne et al., 1980). Both Sephadex G-200-purified seed lectins (lectin 1 and lectin 2) agglutinated R. leguminosarum as did one of two glucose-binding proteins isolated from 7-day-old pea roots which had the same electrophoretic mobility as lectin 2. Host lectin-mediated agglutination of R. leguminosarum and of its purified LPS was inhibited by D-glucose, D-mannose, D-fructose, L-sorbose, maltose, D-melizitose, 2-deoxy-D-glucose, 1-O-methyl-D-glucoside, N-acetyl-D-glucosamine, and 3-O-methyl-D-glucose but not by D-galactose, D-ribose, L-arabinose, D-xylose, D-cellobiose, lactose, raffinose, 6-deoxy-D-glucose, L-rhamnose, 2-deoxy-D-galactose, 2-deoxy-D-glucose-6-phosphate, D-glucoheptose, D-galacturonic acid, D-glucuronic acid, and soluble startch (Kijne et al., 1980).

Information on lectin involvement in host-symbiont recognition among hosts of the cowpea group is scarce. Concavalin A shares immunological cross-reactions with some members of the cowpea group (Hankins et al., 1979). The few reports of experiments on binding of concavalin A with rhizobia and purified EPS or LPS (Dazzo and Hubbell, 1975b; Chen and Phillips, 1976; Kamberger, 1979; Wong, 1980), however, indicate that the binding characteristics displayed

by the lectin of <u>Canavalia ensiforme</u> does not account for the specificity of the <u>Rhizobium</u>-jackbean recognition. Concavalin A binds
with surface determinants present in many rhizobia, regardless of
their ability to nodulate jackbean (Chen and Phillips, 1976; Dazzo
and Hubbell, 1975a; Wong, 1980). The rhizobial binding is inhibited
by α-methyl-mannoside (Dazzo and Hubbell, 1975a), glucose, mannose,
or fructose (Wong, 1980). Specific precipitin lines (1-O-methyl-
L-D-mannopyranoside inhibited) are found in double-diffusion assays
when concavalin A is tested against EPS of <u>Rhizobium</u> 32H1, <u>R.
meliloti</u> strains 2011 and MVH/1, <u>R. leguminosarum</u> LVH/1, or <u>R.
phaseoli</u> DSM30137 and against LPS of <u>R. meliloti</u> MVH/1, <u>R. legum-
inosarum</u> LVH/1, <u>Rhizobium</u> 32H1, <u>R. lupini</u> strains WU425 and H13-3,
or <u>R. japonicum</u> CB1809 (Kamberger, 1979).

   Some literature is available in regard to peanut lectin (PNL)
(Bowles et al., 1979; Pueppke, 1979), also referred to as peanut
agglutinin (PNA) (Bhagwat and Thomas, 1980), and its possible role
in <u>Rhizobium</u> recognition (Bhagwat and Thomas, 1980; Pueppke et al.,
1980; Shulz and Pueppke, 1978). Isoelectric-focusing analysis of
purified preparations of seed lectin of <u>Arachis hypogaea</u> varieties
Spanish and Jumbo Virginia revealed 6 and 7 isolectins, respectively.
Small amounts of lectins were found in 3-day-old roots, but they
were not detectable 9 to 12 days after germination (Pueppke, 1979).
Bowles et al. (1979), however, reported the presence of PNL in the
roots of 35-day-old seedlings. FITC-labelled PNL did not bind to
the cells of 8 strains of <u>Rhizobium</u> nor to 4 strains of <u>R. japonicum</u>
which nodulated <u>A. hypogaea</u> cultivar Jumbo Virginia (Pueppke et al.,
1980; Schulz and Pueppke, 1978). On the other hand, purified FITC-
labelled lectin (FITC-PNA) and ^{125}I-bound PNA were found to adhere
to <u>Rhizobium</u> B.TG-3 and to <u>Rhizobium</u> 5a, which nodulate peanut, but
not to <u>R. japonicum</u> or <u>R. meliloti,</u> which do not nodulate peanut.
The results point to rhizobial EPS as the major sites for lectin,
and to LPS as the secondary binding sites (Bhagwat and Thomas, 1980).
In conclusion, the available data do not favor the implication of
PNL in peanut-<u>Rhizobium</u> recognition.

   Hamblin and Kent (1973) reported the binding of seed phyto-
hemagglutinin of <u>Phaseolus vulgaris</u> to <u>R. phaseoli</u> cells. Experiments with ^{32}P-labelled <u>Rhizobium</u> have shown that <u>R. phaseoli</u>, among
other <u>Rhizobium</u> species studied, binds to pea roots (Chen and
Phillips, 1976). Nonspecific binding of FITC-labelled phytohemaglu-
tinin (<u>Phaseolus vulgaris</u> lectin) was observed (Chen and Phillips,
1976; Law and Strijdom, 1977). Binding of FITC-phytohemaglutinin
to the homologous nodulating <u>R. phaseoli</u> was not recorded for strain
127K17 (Chen and Phillips, 1976) and for 4 of 10 strains tested
(Law and Strijdom, 1977). Moreover, the binding of FITC-phyto-
hemagglutinin to 3 of 7 (Chen and Phillips, 1976) and to 3 of 4 (Law
and Strijdom, 1977) heterologous <u>Rhizobium</u> strains has been demonstrated by fluorescence microscopy. The results on the interactions
of lectins with components of the rhizobial cell surface are con-

flicting. Specific interactions of R. phaseoli LPS with immobilized
lectin of red kidney have been described (Albersheim et al., 1977;
Wolpert and Albersheim, 1976), but no correlation between LPS structure and nodulation groups (including R. phaseoli) has been suggested
(Carlson et al., 1978; Zevenhuizen et al., 1980). Furthermore, purified R. phaseoli DSM30137 LPS and EPS show no specific interaction
with purified P. vulgaris lectin (Kamberger, 1979). Most of the
published evidence herein described shows no obvious specific relation of P. vulgaris lectin and R. phaseoli in host-symbiont recognition.

     In regard to the R. meliloti-medic system, the binding of R.
meliloti T3T-conjugated capsular polysaccharides to alfalfa root
hairs but not to clover roots has been reported (Dazzo et al., 1976;
Dazzo and Brill, 1977); however, a role for LPS (but not EPS) of
R. meliloti strains 2011 and MVH/1 in host-symbiont recognition has
also been suggested on the basis of specific precipitin reactions
between purified LPS and lectin purified from Medicago sativa seeds
(Kamberger, 1979). This reaction is inhibited by 0.2 N-acetylgalactosamine.

     Little is known of the R. lupini-lupin association in regard
to host symbiont. Reports on the binding of R. lupini cells to
heterologous hosts (like pea) and to FITC-lectins (Chen and Phillips,
1976) as well as the interaction of purified EPS and LPS with purified heterologous legume lectins (Kamberger, 1979) do not allow for
any generalizations.

FINAL REMARKS

     Enough evidence has been published on clover-R. trifolii inter-
actions in regard to the host-symbiont recognition phenomenon to
support the lectin-recognition hypothesis suggested by Bohlool and
Schmidt (1974) and the postulated cross-bridging model in which
cross-reactive antigens present on the surface of both the host
(root hairs) and the symbiont (EPS) bind clover lectins (Dazzo and
Hubbell, 1975b). Some important questions, such as how does host-
symbiont recognition occur in soybean lectin SBL, still need to be
answered in order to fit experimental observations with the lectin-
recognition hypothesis in the soybean-R. japonicum system. Additional questions are yet to be answered in regard to the remaining
host-Rhizobium systems.

     It is possible that more information on the legume root glyco-
proteins that bind Rhizobium cells will shed additional light on the
recognition phenomenon. Erythrocytes agglutinating seed lectins
(phytohemagglutinins) have been preferentially used to test the
lectin-recognition hypothesis. However, there is no reason to believe that these particular glycoproteins are necessarily the only

ones that may be involved in host-symbiont recognition. Even if
the properties of a particular seed lectin can account for the
specific adsorption of homologous Rhizobium, it is necessary to show
the presence of such a lectin on a particular root epidermal region
where the infection process will take place (Dazzo et al., 1978).
The presence of glycoproteins other than seed lectins has been reported for some legume roots; e.g., soybeans (Kagestra and Andrews,
1978) and peas (Kijne et al., 1980). The role of lectins in host-
symbiont recognition may involve more than serving as a mere micro-
symbiont-binding element. Kijne et al. (1980) suggested that pea
lectins may be involved in the metabolism of glycans in the young
lectinic primary cell wall assembly, and enzymatic functions have
been reported for legume lectins (Albersheim and Wolpert, 1976;
Hankins and Shannon, 1978; Hankins et al., 1979).

    Rhizobial EPS and LPS have been said to contain the receptor
site(s) for specific attachment to the host (Dazzo and Hubbell,
1975b; Kamberger, 1979). Acidic EPS's are the major receptor sites
for trifoliin (Dazzo and Hubbell, 1975b). However, a secondary
binding site is associated with the O antigen portion of LPS (Dazzo
and Brill, 1979). Similar results have been reported for PNL-
Rhizobium binding (Bhagwat and Thomas, 1980). The suggestion of
dual binding sites on the surface of rhizobial cells seems quite
plausible since production of EPS relies on the amount of carbohydrate available. In the soil, under normal conditions, limiting
amounts of carbohydrates may result in poor synthesis of EPS. At
such times, the secondary lectin receptor located within the LPS
structure may become of utmost importance for the establishment of
the symbiosis.

REFERENCES

P. Albersheim, A. R. Ayers, Jr., B. S. Valent, J. Ebel, M. Hahn,
     J. Wolpert, and R. Carlson, J. Supramol. Struct., 6:599-616
     (1977).
P. Albersheim and J. S. Wolpert, Plant Physiol. Suppl., 58:79
     (1976).
A. K. Bal, S. Shantharam, and S. Ratnman, J. Bacteriol., 133:1393-
     1400 (1978).
A. A. Bhagwat and J. Thomas, J. Gen. Microbiol., 117:119-125 (1980).
T. V. Bhuvaneswari and W. D. Bauer, Plant Physiol., 62:71-74 (1978).
T. V. Bhuvaneswari, S. G. Pueppke, and W. D. Bauer, Plant Physiol.,
     60:486-491 (1977).
P. E. Bishop, F. B. Dazzo, E. R. Appelbaum, R. J. Maier, and W. J.
     Brill, Science, 198:938-940 (1977).
B. B. Bohlool and E. L. Schmidt, Science, 185:269-271 (1974).
B. B. Bohlool and E. L. Schmidt, J. Bacteriol., 125:1188-1194 (1976).
P. M. Bonish, N. Z. J. Agric. Res., 23:239-242 (1980).

D. J. Bowles, H. Lis, N. Sharon, Planta, 145:193–198 (1979).
J. Brockwell and F. W. Hely, Aust. J. Agric. Res., 17:884–899
    (1966).
R. E. Buchanan and N. E. Gibbons, Bergey's Manual of Determinative
    Bacteriology, Williams and Wilkins, Baltimore (1974).
B. E. Caldwell and G. Vest, Crop Sci., 8:680–682 (1968).
B. E. Caldwell and G. Vest, Crop Sci., 10:19–21 (1970).
H. E. Calvert, M. Lalonde, T. B. Bhuvaneswari, and W. D. Bauer,
    Can. J. Microbiol., 24:785–793 (1978).
R. W. Carlson, R. E. Sanders, C. Napoli, and P. Albersheim, Plant
    Physiol., 62:912–917 (1978).
N. Catsimpoolas and E. W. Meyer, Arch. Biochem. Biophys., 132:279–
    285 (1979).
A. T. Chen and D. A. Phillips, Physiol. Plant., 38:83–88 (1976).
R. A. Date, Soil Biol. Biochem., 5:5–18 (1973).
F. B. Dazzo and W. J. Brill, Appl. Environ. Microbiol., 33:132–136
    (1977).
F. B. Dazzo and W. J. Brill, Plant Physiol., 62:18–21 (1978).
F. B. Dazzo and W. J. Brill, J. Bacteriol., 137:1362–1373 (1979).
F. B. Dazzo and D. H. Hubbell, Plant Soil, 43:713–717 (1975a).
F. B. Dazzo and D. H. Hubbell, Appl. Microbiol., 30:1017–1033
    (1975b).
F. B. Dazzo, C. A. Napoli, and D. H. Hubbell, Appl. Environ. Micro-
    biol., 32:166–171 (1976).
F. B. Dazzo, M. R. Urbano, and W. J. Brill, Curr. Microbiol., 2:15–
    20 (1979).
F. B. Dazzo, W. E. Yanke, and W. J. Brill, Biochim. Biophys. Acta,
    539:276–286 (1978).
J. de Ley, Annu. Rev. Phytopathol., 6:63–90 (1968).
W. F. Dudman, Carbohydr. Res., 46:97–110 (1976).
W. F. Dudman, Carbohydr. Res., 66:9–23 (1978).
D. W. Fountain and W. Yang, Biochim. Biophys. Acta, 492:176–185
    (1977).
A. H. Gibson, Aust. J. Agric. Res., 19:907–918
A. H. Gibson and J. Brockwell, Aust. J. Agric. Res., 19:891–905
    (1968).
P. Graham, in: Symbiotic Nitrogen Fixation in Plants (P. S. Nutman,
    ed.), Cambridge University Press, London, pp. 99–118 (1975).
P. H. Graham and D. H. Hubbell, in: Forage in Livestock Production,
    Doll and Nott, Special Publication No. 24, American Society of
    Agronomy, Madison, Wisconsin, pp. 9–21 (1975).
J. Hamblin and S. P. Kent, Nature New Biol., 245:28–30 (1973).
C. N. Hankins, J. I. Kindinger, and L. M. Shannon, Plant Physiol.,
    64:104–107 (1979).
C. N. Hankins and L. M. Shannon, J. Biol. Chem., 253:7791–7797
    (1978).
C. M. Hepper, Ann. Bot., 42:109–115 (1978).
C. M. Hepper and L. Lee, Plant Soil, 51:441–445 (1979).
T. Hymowitz, Econ. Bot., 24:408–421 (1970).

W. Kamberger, Arch. Microbiol., 121:83-90 (1979).

G. Kato, Y. Maruyama, and M. Nakamura, Agric. Biol. Chem., 43:1085-1092 (1979).

K. Keegestra and J. Andrews, Plant Physiol. Suppl., 61:19 (1978).

J. W. Kijne, I. A. M. Van Der Schaal, and G. E. De Vries, Plant Sci. Lett., 18:65-74 (1980).

R. T. Lange, in: Bacterial Symbiosis with Plants, Vol. 1 (S. M. Henry, ed.), Academic Press, New York, pp. 99-170 (1966).

I. J. Law and B. W. Strijdom, Soil Biol. Biochem., 9:79-84 (1977).

H. Lis, C. Friedman, N. Sharon, and E. Katchalski, Arch. Biochem. Biophys., 117:301-309 (1966).

R. J. Maier, P. E. Bishop, and W. J. Brill, J. Bacteriol., 134:1199-1201 (1978).

R. J. Maier and W. J. Brill, J. Bacteriol., 133:1295-1299 (1978).

A. J. Mort and W. D. Bauer, Plant Physiol., 66:158-163 (1980).

C. Napoli and P. Albersheim, J. Bacteriol., 141:979-980 (1980a).

C. Napoli and P. Albersheim, J. Bacteriol., 141:1454-1456 (1980b).

D. O. Norris and L. t'Mannetje, E. Afr. Agric. For. J., 29:214-235 (1964).

J. H. Orf, T. Hymowitz, S. P. Pull, and S. G. Pueppke, Crop Sci., 18:899-900 (1978).

K. Planque and J. W. Kijne, FEBS Lett., 73:63-66 (1977).

S. G. Pueppke, PLant Physiol., 64:575-580 (1979).

S. G. Pueppke, W. D. Bauer, K. Keegstra, and A. L. Ferguson, Plant Physiol., 61:779-784 (1978).

S. G. Pueppke, T. G. Freund, B. C. Schulz, and H. P. Friedman, Can. J. Microbiol., 26:1489-1497 (1980).

S. P. Pull, S. G. Pueppke, T. Hymowitz, and J. H. Orf, Science, 200:1272-1279 (1978).

B. K. Robertson, P. Aman, A. G. Darvill, M. McNeil, and P. Albersheim, Plant Physiol., 67:389-400 (1981).

B. G. Rolfe, P. M. Gresshoff, J. Shine, and J. M. Vincent, Appl. Environ. Microbiol., 39:449-452 (1980).

R. E. Sanders, R. W. Carlson, and P. Albersheim, Nature (London), 271:240-242 (1978).

B. C. Schulz and S. G. Pueppke, Plant Physiol. Supppl., 61:59 (1978).

S. Shantaharam, J. A. Gow, and A. K. Bal, Can. J. Microbiol., 26:107-114 (1980).

G. S. Stacey, A. S. Paau, and W. J. Brill, Plant Physiol., 60:609-614 (1980).

R. W. Stahlhut and T. Hymowitz, Soybean Genet. News, 7:41-43 (1980).

R. W. Stahlhut, T. Hymowitz, and J. H. Orf, Crop Sci., 21:110-112 (1981).

L. Su, S. G. Pueppke, and H. P. Friedman, Biochim. Biophys. Acta, 629:292-304 (1980).

L. t'Mannetje, Antonie van Leewenhoek J. Microbiol. Serol., 33:477-491 (1967).

M. J. Trinick, Aust. J. Sci., 27:263-264 (1964).

M. J. Trinick, Exp. Agric., 4:243-253 (1968).

H. C. Tsien and E. L. Schmidt, Can. J. Microbiol., 23:1274–1284
    (1977).
H. C. Tsien and E. L. Schmidt, Appl. Environ. Microbiol., 39:1100–
    1104 (1980).
H. C. Tsien and E. L. Schmidt, J. Bacteriol., 145:1063–1074 (1981).
J. M. Vincent, in: Nitrogen Fixation. Vol. 2, Symbiotic Associa-
    tion and Cyanobacteria (W. E. Newton and W. H. Horme-Johnson,
    eds.), University Park Press, Baltimore, pp. 103–109 (1980).
J. S. Wolpert and P. Albersheim, Biochem. Biophys. Res. Commun.,
    70:729–737 (1976).
P. P. Wong, Plant Physiol., 65:1049–1052 (1980).
L. P. T. M. Zevenhuizen, I. Scholten-Koerselman, and M. A. Posthumus,
    Arch. Microbiol., 125:1–8 (1980).

GENETICS AND BREEDING OF NITROGEN FIXATION

Thomas E. Devine

Plant Physiology Institute
U.S. Department of Agriculture
Beltsville, Maryland 20705. U.S.A.

Problems and Questions

Frequently during the introduction of a legume crop to a new agricultural environment, the simultaneous introduction of the appropriate rhizobial microsymbiont results in a marked and clearly evident enhancement of plant growth and crop yield. After this initial success has been achieved, the question arises, "What more can be done to improve crop performance through nitrogen fixation?" There are many important yet unresolved questions related to this central question. For example: Is nitrogen availability a factor limiting crop yield? Is the efficiency of nitrogen fixation a factor limiting crop yield? Can the total amount of nitrogen fixed be increased? Can the efficiency of nitrogen fixation be improved? If so, how? Can the genetic control system of the host plant which governs nitrogen fixation be altered advantageously? Can the genetic control system of the rhizobial microsymbiont be modified to improve agronomic performance? Can the rhizobial populations in the field be specified? Can the rhizobial population admitted to symbiosis be specified? Can the mutual compatibility of host and microsymbiont be enhanced? If nitrogen fixation is increased or improved in efficiency, are other correlated processes in plant growth or crop performance also altered? Does improved symbiotic fixation reduce carbon available for seed storage or root growth? Is disease resistance affected? Are insect or nematode resistance altered? Is maturity affected? Is plant lodging affected? What are the costs of the resources invested in breeding for enhanced nitrogen fixation? Is selection for enhanced nitrogen fixation compatible with other objectives in a crop breeding program? Will resources have to be diverted from other breeding objectives? Relative to other breeding objectives (such as disease and insect resistance, tolerance of edaphic factors, tolerance of temperature and water stress, harvest index, etc.), what degree of improvement can be expected from nitrogen fixation and with what probability of success?

I am convinced that the fundamental question of "Whether or not the enhancement of biological nitrogen fixation will increase crop yields?" will not be resolved by deductive "a priori" reasoning. The proposition must be tested empirically by developing (perhaps through several breeding strategies) plant populations bred for improved nitrogen fixation. The actual performance of such populations under field conditions will then be used to determine both the effective realized heritability of nitrogen fixation and the efficacy of such selection for improving crop yield and agronomic performance.

A Complex Challenge

The challenge of improving symbiotic nitrogen fixation is made especially complex because we are dealing with two genetic systems of phylogenetically disparate organisms: (a) the plant macrosymbiont and (b) the bacterial microsymbiont, the Rhizobium. There is no track record of performance for this endeavor, nor is there precedent to guide the enterprise. Plant breeding has been eminently successful in improving the production of many crops through selection for traits such as lodging resistance, maturity, shattering resistance, winter hardiness, disease, and insect and nematode resistance. Although breeding for resistance to pathogens involves both the host and microorganisms, the relationship and goal are radically different from breeding for enhanced nitrogen fixation. In the case of breeding for disease resistance, the goal is to separate the host from the pathogen - to interrupt the relationship. In contrast, breeding to enhance symbiotic nitrogen fixation requires preserving the relationship and improving the metabolic integration. Success is by no means assured, but there is a reasonable basis for hope. Nature has selected for survival ability, not for agronomic productivity. Therefore, there should be opportunity for improvement in agronomic performance.

UNDERSTANDING THE GENETIC AND BIOLOGICAL NATURE OF SYMBIOSIS

Genetic Inventory for the Soybean Host

That symbiosis is under genetic control is well established. Let us consider the inventory of genetic factors controlling symbiosis in soybeans (Table 1). Several host-plant genes that control interaction with the microsymbiont (designated Rj genes to refer to the microsymbiont Rhizobium japonicum) have been reported. The rj_1 gene, a recessive gene, produces the so called "nonnodulating"

TABLE 1. Host Genetic Factors Controlling Symbiosis in Soybeans

Genetic factors	Soybean strain	Phenotype
rj_1	T181, T201	Nonnodulating with most strains
Rj_2	Hardee, CNS	Ineffective with strains of Cl and 122 serogroups
Rj_3	Hardee	Ineffective with strain 33
Rj_4	Hill, Dunfield	Ineffective with strain 61
?	Peking	Ineffective with strains of serogroup 123

phenotype with a broad spectrum of Rhizobium strains. The gene was discovered by Williams and Lynch (1954) in a single line of parents with normal nodulation behavior. There have been no other independent reports of its occurrence. It appears to be a unique mutational event. In contrast, three other genes, Rj_2, Rj_3, and Rj_4 (Caldwell, 1966; Vest, 1970; Vest et al., 1972), are all dominant genes. These alleles are reported to produce an ineffective nodulation response with either a serological group of strains or a single strain of Rhizobium. The ineffective reaction is characterized by the development of cortical proliferations on the roots or rudimentary nodules rather than normally developed nodules. Another form of interaction is seen in the case of the soybean cultivar "Peking" inoculated with Rhizobium strain 123 of the Beltsville Culture Collection (Vest et al., 1973). In this case, an ample number of nodules of normal size are developed, but nitrogen fixation is virtually nil. The nodule interior lacks the pink color characteristic of tissue containing leghemoglobin. The Peking soybean with Rhizobium 123 combination is an extreme example of inefficiency in nitrogen fixation. Examples of highly efficient combinations are many. Rhizobium strains 110 and 142 of the Beltsville Culture Collection are highly efficient in combination with many United States-bred soybean cultivars. Between these extremes, other combinations cover a wide range in the efficiency of nitrogen fixation.

Significance of Host-Incompatibility Alleles

The question may be posed, "Why do we find these incompatible reactions?" The suggestion immediately occurs that the Rj alleles are mutations - inborn metabolic errors - errors in replication of the DNA that have been perpetuated in our breeding lines. However, most mutations are recessive, and whereas this may be the case with rj_1, it is not the case with the Rj_2, Rj_3, and Rj_4 alleles. This incongruity of the mutation hypothesis with the dominant character

TABLE 2.   Summary of Frequencies of Rj_2 and Rj_4 Alleles in Soybean
           Plant Introduction Lines by Maturity Group and Country
           of Origin

Category	Plant introduction lines (No.)	Frequency of alles (%)	
		Rj_2	Rj_4
Maturity group			
00	11	0	9.0
0	23	0	15.8
I	94	0	12.8
II	93	0	16.1
III	93	1.1	25.8
IV	93	1.1	33.3
V	98	1.0	26.5
VI	98	4.1	37.9
VII	90	7.1	27.5
VIII	99	3.4	39.3
IX	20	0	24.7
X	35	0	94.3
Total or proportion of total	847	2.0	29.7
Country of origin			
U.S.S.R	14	0	0
China	248	4.0	35.7
Korea	171	.6	19.4
Japan	243	2.2	24.2
Taiwan	5	0	60.0
Vietnam	5	0	97.4
Burma and Malaya	14	0	71.7
Thailand	26	0	66.9
Indonesia	33	1.0	64.8
India and Pakistan	88	0	15.7
Total or proportion of total	847	2.0	29.7

of the Rj_2, Rj_3, and Rj_4 alleles leads us to look further for an
explanation. An alternative explanation may be derived from a consideration of the history of soybean production (Devine and Breithaupt, 1980c, 1981).

Soybeans, and presumably their specific microsymbiotic rhizobia,
were introduced into the United States from Asia. Some cowpea rhizobia, however, not necessarily of Asiatic origin, may nodulate soybeans. It is reasonable to postulate that within particular eco-

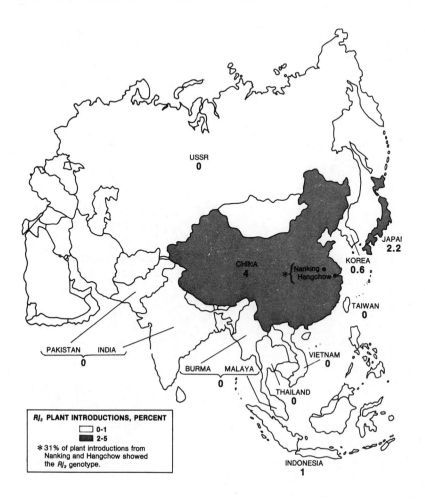

Fig. 1.   Frequency of Rj₂ alleles.

logical areas in Asia, a natural selection pressure was exerted upon
both the host and microsymbiont for mutual compatibility in symbiosis. Thus, mutually adapted and compatible ecotypes of the host
and microsymbiont would evolve. When ecotypes of the host and microsymbiont, that have not coevolved, are artificially brought into association (either in laboratory experiments or in field used for crop
production), the symbiosis may be defective in either the effectiveness of nodulation, the efficiency of fixation, or both. The Rj_2,
Rj_3, and Rj_4 responses may then be indicative of the association of
ecotypes of the host that have not coevolved with ecotypes of the
microsymbiont. For example, during soybean introduction into the
United States, the reassortment of ecotypes of the host and micro-

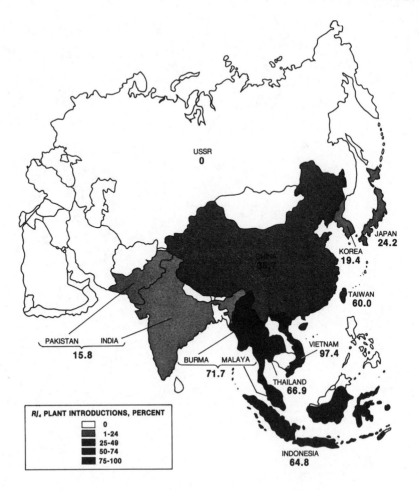

Fig. 2.   Frequency of Rj₄ alleles.

symbiont may have resulted in the coupling of ecotypes of <u>R. japoni-</u>
<u>cum</u> from northern Asia with soybean host-plant germplasm from
southern Asia.

To test the validity of these hypotheses, we surveyed a sample
of the Asiatic soybean lines from a range of different locations and
different maturity groups for the presence of the Rj₂ and Rj₄ alleles.
If the Rj₂ and Rj₄ alleles were simply a mutation phenomenon, then
we would expect the frequency of these alleles to be low and their
occurrence to be random with respect to geographic distribution. On
the other hand, if the incompatible reactions result from coupling

genotypes of the host and microsymbiont that have not coevolved in a common ecological range, then the frequencies of the Rj_2 and Rj_4 alleles need not be low and a geographic pattern to their distribution should be detectable. A total of 847 plant introduction lines (PI's) were tested in growth trays (Devine and Reisinger, 1978) for the presence of the Rj_2 and Rj_4 phenotype. The results are shown in Table 2 and Figs. 1 and 2. The Rj_2 allele occurred with relatively low frequency 2% of the lines in the sample of 847 PI's. These lines were in maturity groups III through VIII, and most originated in China and Japan (Table 2, Fig. 1). There appears to be a clustering of the Rj_2 allelic frequency in the vicinity of the neighboring Chinese cities of Nanking and Hangchow, which are 240 km apart. Of the total 18 PI's found to carry the Rj_2 allele in this testing program, 9 were from these two cities. Furthermore, of the 29 lines tested from these cities, 9 (or 31%) displayed the Rj_2 allele. Although the sample of 29 lines is small, the frequency of 31% is markedly different from the overall frequency of 2% in the total of all lines.

The Rj_4 allele occurred in 29.7% of the 847 PI's tested for this allele. It was found in all the maturity groups but with a lower frequency in the earlier groups, 00-II, than in the later ones (Table 1). The highest frequency (94.3%) was in maturity group X. The Rj_4 allele was particularly prevalent in lines from southeast Asia, i.e., from Burma, Malaya, Indonesia, Thailand, and Vietnam. Only a few lines are available in the U.S. Department of Agriculture soybean germplasm collections for some of these countries, i.e., Vietnam 5, Burma 2, and Malaya 12; however, the frequencies are consistent with those of neighboring countries (Fig. 2). These results indicate that there is a relationship between the geographical sources of the PI's and the frequencies of the Rj alleles.

If host compatibility with the microsymbiont is related to ecotype coevolution, and if successful symbiotic nitrogen fixation is an important component of the complex of adaptive characteristics for host survival, then it would appear that the Rhizobium strains with those symbiotic characteristics recognized as incompatible by the Rj_2 and Rj_4 alleles do not dominate the rhizobial population (so as to exert a significant selection pressure) in areas where the host populations have a high frequency of the Rj_2 and Rj_4 alleles.

The Rj genes produce phenotypes that are strikingly visible evidence of host-microsymbiont incompatibility. There are, however, more subtle, less easily distinguishable deficiencies in the efficiency of symbiosis following nodule formation. If the same principle of ecotypic coevolution that governs the effects of the Rj_2 and Rj_4 alleles on effectiveness of nodulation also governs the efficiency of nitrogen fixation, then coupling the proper ecotypes of the soybean host and rhizobial microsymbiont should result in improved efficiency of nitrogen fixation.

To understand the role of ecotype coevolution in the development of symbiotic specificity, it will be important to identify precisely the origin of both host and microsymbiont genotypes. The information now available on soybean PI's permits such identification of many lines. Unfortunately, we have only meager information on most strains of R. japonicum. Hopefully, new introductions of both R. japonicum and soybeans will be definitively described. If research can identify ecotypes of Rhizobium that show superior combining ability with specified host germplasm, it should prove possible to successfully predict the locations from which new rhizobial collections are most likely to yield the most compatible strains.

To decipher the ecotypic relationships, it will be necessary to collect simultaneously both the host and associated rhizobia. The usual and most convenient method of collecting the host is to collect the seed. Unfortunately, in an annual species such as soybeans, by the time the seed is ripe, the nodules on the roots have degenerated, making it difficult to obtain uncontaminated rhizobial cultures. This problem might be resolved by collecting the seed and at the same time collecting soil from the plant site. Later, the appropriate seed could be planted in this soil in a greenhouse, and cultures would be obtained from the resulting nodules. Unfortunately, in the case of soybeans and its associated microsymbiont, this task has not been adjudged of sufficiently high priority to have been initiated. The widening use of modern inoculation techniques will likely result in a mixture of rhizobial genotypes in soybean production fields in Asia. Ths will soon make it impossible to detect the ecotypic relationships.

The situation with peanuts is much brighter. The University of North Carolina at Raleigh holds extensive, well documented collections of both host and microsymbiont germplasm.

An understanding of ecotype variation may provide insight into the occurrence of types of Rhizobium with differing physiological characteristics, such as rhizobial-induced chlorosis (Johnson and Means, 1960), hydrogenase activity (Schubert and Evans, 1976), and tolerance of extremes of pH (Ham et al., 1971) and temperature (Munevar and Wollum, 1980). Such differences may have survival value in specific ecological niches. An understanding of these adaptive functions would benefit efforts to tailor combinations of host plants with desirable Rhizobium strains for agricultural ecosystems.

Differentiation of Ecotypic Affinities

Evidence for intraspecific differentiation of host affinities for populations of Rhizobium strains may be gleaned from the reactions of the soybean cultivar Peking. As mentioned earlier, when

Peking is inoculated with Rhizobium strain 123, nodulation occurs,
but the nodules formed fix virtually no nitrogen. Further, Peking
exhibits a strong negative preference against Rhizobium strain 110
of the Beltsville Culture Collection (Caldwell and Vest, 1968).
Strain 110 is an excellent nitrogen fixer with most United States-
bred soybean cultivars. When it is used as the sole inoculum source
on Peking, it results in nodulation and nitrogen fixation. However,
when Peking is planted in soil containing a mixture of strain 110
and other strains, very few nodules containing strain 110 are formed.
In addition, the recently discovered "fast growing" Rhizobium strains
(Keyser et al., 1981a), which are capable of nodulating and fixing
nitrogen with some soybeans, will nodulate Peking but fail to form
a successful symbiosis with several United States-bred cultivars.
I would interpret the evidence presented above as indicating that
Peking differs from the predominant germplasm used in current United
States-bred soybean cultivars with respect to a combination of char-
acteristics related to symbiotic affinity. Peking is probably not
unique, but it is representative of a soybean subpopulation.

There is considerable interest in the hydrogen metabolism of
rhizobia in relation to the efficiency of fixation (Schubert and
Evans, 1976; Evans et al., 1981; Hanus et al., 1981). It has been
suggested, on the basis of chemical kinetics, that Hup^+ strains
of Rhizobium with uptake hydrogenase should be more efficient in
nitrogen fixation than those without this system. Recently, research at Beltsville has indicated that some Rhizobium strains,
when tested on soybeans, lack the uptake hydrogenase system (Hup^-)
and are relatively mediocre in fixing nitrogen, but when tested on
cowpeas, these strains exhibit the hydrogenase system (Hup^+) and
are very active in fixing nitrogen. Thus, the expression of the
hydrogenase phenotype depends not only on the genotype of the rhizobia, but also on the genotype and the species of the host plant.
Rhizobium strains previously thought to be inefficient or poorly
performing strains were not, in fact, inherently poor, but were
merely not being evaluated on the proper host.

Evidence of ecotypic variation in rhizobia may be inferred from
a survey by Ham et al. (1971) of isolates from soybean nodules from
75 locations in Iowa. Strains of Rhizobium of serogroup 123 were
found to be the most numerous group in soils with a pH below 7.5,
but in soils with a pH above 7.8, strains of serogroup 135 were dominant in the population. Evidence that different ecological types
of clover rhizobia have differential compatibility with host ecotypes can be found in the research of Masterson and Sherwood (1974).
They reported that ecotypes of rhizobia from wet, acid soils in
Ireland were compatible with local ecotypes of clover adapted to
such sites. However, these same rhizobia were often ineffective with
the agronomically bred clover, "S100." Beringer et al. (1980) summarized the evidence suggesting that rhizobial plasmids carry genetic material determining host-range specificity.

From the evidence presented above (i.e., that there are intra-specific differences in host affinity for rhizobial populations as with Peking, and the contingency of Rhizobium strain performance on host genotype), I suggest the general principle that the Rhizobium strains found in nature are almost always both effective and effi-cient, and that these strains would be recognized as such if tested on their appropriate host genotype. Therefore, the lack of expres-sion of the hydrogenase system in symbiosis with soybeans by some strains considered to be members of the species R. japonicum may arise because they are not, in fact, evolutionarily differentiated as homologs of soybeans.

It is a most important, but, unfortunately, frequently neglec-ted fact that symbiotic nitrogen fixation is a biological process. The biological dimension embraces the concept of value - biological processes have a value in relation to the survival of an organism. Biological processes are the product of evolutionary development, in which natural selection acts to develop and adapt the process to the total and particular environment experienced by the breeding popu-lation. There is a tendency to remove consideration of symbiotic nitrogen fixation from its biological dimension and view it only as a chemical process. Such an approach is not adequate.

Effect of Agronomic Selection
on Host-Incompatibility Alleles

We are interested in knowing whether selection for agronomic improvement of soybeans in the breeding programs conducted in the United States has resulted in a shift in the frequencies of genes affecting nitrogen fixation (Devine and Breithaupt, 1981). To de-termine this, we compared the frequencies of the Rj_2 and Rj_4 alleles in the PI lines and the 1979 preliminary and uniform test lines to determine whether a shift in the frequencies of these alleles oc-curred concomitantly with selection for improved agronomic perfor-mance (Table 3). The PI's are the raw material with which plant breeders initiate plant improvement programs. The preliminary test lines represent the products of recombination and selection that have performed sufficiently well agronomically in local or statewide testing programs to merit entry in the national testing program. The uniform test lines represent a yet more advanced level of se-lection, restricted to those that have previously performed well in the preliminary tests.

To strengthen the basis for comparison, we selected the PI's acquired at the earliest dates for testing. Lines in the 1979 pre-liminary and uniform tests are usually derived from crosses made 10 or more years previously. Thus, the PI's acquired at the earlier dates are more apt to represent the germplasm available for utiliza-tion as parental materials in those crosses. Although relatively few lines served as the source of our currently used germplasm (15

TABLE 3. Comparison of Frequencies of Rj_2 and Rj_4 Alleles in Soybean Plant Introduction Lines and 1979 United States Regional Preliminary and Uniform Soybean Test Lines

Allele	Line	Number	Percent
Rj_2	Plant introduction lines	758	2.2
	Preliminary test lines	260	0.4
	Uniform test lines	103	0.0
Rj_4	Plant introduction lines	758	27.6
	Preliminary test lines	260	13.7
	Uniform test lines	103	8.5

PI's are reported to account for approximately 80% of the currently used germplasm in the United States), presumably the lines carrying the Rj alleles had a proportional opportunity to be selected as parents. Selection was practiced by breeders for agronomic desirability, both in selecting parental materials and in culling the progeny of crosses among lines selected for use as parents.

The Rj_4 allele occurred with a relatively high frequency in the plant introductions (29.7%) and should therefore have had ample opportunity, based on random chance, to have been included in the selection of parental materials. The Rj_2 allele, on the other hand, was relatively infrequent in the plant introductions (2%); however, the assignment of cultivar names to several of the PI's carrying the Rj_2 allele indicates that these lines were deemed to have some features of agronomic merit and should, therefore, at least have been considered in the selection of parental materials. Until recently, the genetic constitution at the Rj_2 and Rj_4 loci would not have been known to breeders for most soybean lines. Consequently, direct conscious selection for these alleles would not have been possible.

Data for the preliminary and uniform tests do not represent a sample but are the entire population of experimental lines in these tests in 1979. In the case of both the Rj_2 and Rj_4 alleles, the frequencies of the incompatible alleles progressively declined from the PI's to the preliminary tests to the uniform tests (Table 3).

Although a shift in allelic frequency can occur because of random drift, particularly in small populations, it seems less likely that this would account for the consistent trend with both alleles. The alternative hypothesis would be that lines carrying either Rj_2 or Rj_4 may have been generally substandard performers agronomically, and, thus, have been eliminated in the culling process leading to advancement to the 1979 preliminary and uniform tests. Even so, some cultivars known for good agronomic performance do carry the Rj_2 or Rj_4 allele.

It is not clear whether genotypes are disadvantaged in performance in United States testing programs because of the close linkage of the Rj_2 or Rj_4 alleles to other detrimental alleles or because of an inherent adaptive deficiency of the Rj_2 and Rj_4 alleles themselves. The makeup of the rhizobial populations in the evaluation sites used for soybean testing in the United States is not well understood. Possibly at some locations, the Rj_2 and Rj_4 alleles may significantly impair nitrogen fixation and are therefore detrimental to performance. It would be desirable to have more information on the rhizobial populations at the testing sites used in breeding programs.

Hydrogenase Activity and Compatibility

Dixon (1972) reported that the expression of hydrogenase activity by a Rhizobium strain depended on the host species. Strain ONA311 exhibited different amounts of hydrogenase activity with the host species Pisum sativum and Vicia bengalensis, but in nodules of Vicia faba, the enzyme was absent. Gibson et al. (1981) reported that rhizobial strains CB756 and 32H1 produced nodules that lacked hydrogenase (Hup⁻) with the host species Vigna radiata, but expressed hydrogenase (Hup⁺) with V. unguiculata, V. mungo, and two other Vigna species. Keyser et al. (1981b, 1981c) reported that two strains of Rhizobium evolved hydrogen at significantly higher rates in symbiosis with soybeans than with cowpeas.

Nangju (1980) reported that research in field plots in Nigeria showed that the soybean cultivars Malayan, Orba, and TGm 686 from Southeast Asia nodulated and fixed nitrogen with indigenous African rhizobia. However, the American cultivars TGm 294-4-2371, Bossier, and Jupiter nodulated poorly with the indigenous rhizobia. The author postulated that the indigenous rhizobia belong to the cowpea-type rhizobia, which are known to be widely distributed in Africa. A few Rhizobium strains of the cowpea group have been reported to induce nodule formation on some cultivars of soybeans (Leonard, 1923). Nevertheless, R. japonicum has been considered principally as specific for soybeans. The generally accepted definition of R. japonicum has rested on the axiom that rhizobia that nodulate soybeans are R. japonicum. This concept should be re-evaluated.

A survey of the frequency of Rhizobium strains expressing an active hydrogenase system in vitro was conducted in the United States (Lim et al., 1981). The variation in expression of Hup activity in vivo resulting from the host genotype leaves some ambiguity regarding the meaning of in vitro determinations. Cultures were derived from soybean nodules from 70 locations in 28 states representing the areas of soybean production. Most of the isolates (more than 75%) lacked an active hydrogenase system. Especially interesting is the fact that isolates from the southeastern states (Alabama, Florida, Mississippi, North Carolina, and Louisiana) were predominantly Hup⁻ strains.

Cowpeas are thought to have been introduced to the eastern portion of the United States during the colonial period. Successful nitrogen fixation by these plants suggests that the homologous rhizobia were also introduced at this time. Cultivation of cowpeas in the southeastern United States has been common since that period. Several factors suggest to this author that some strains of Rhizobium nodulating soybeans in the United States may, in fact, be only partially homologous with soybeans and may be more nearly homologous with cowpeas or legumes with symbiotic affinities similar to cowpeas. These factors are the following: (a) the expression of hydrogenase activity by the rhizobia depends upon the host genotype; (b) some rhizobia, thought to be R. japonicum, form an effective symbiosis with cowpeas and behave as Hup⁺ strains on cowpeas, but as Hup⁻ on soybeans; (c) soybean genotypes from southeast Asia nodulate with indigenous rhizobia in Africa, now postulated to be cowpea rhizobia; and (d) the history and distribution of cowpea introduction and cultivation in the southeastern United States is congruent with the high incidence of recovery of Hup⁻ rhizobia from the southeastern states.

In the case of the symbiosis of rhizobia with soybeans, cowpeas and some other legumes, we may be dealing with populations still in an active state of evolution in which differentiation of the rhizobia for specialized affinity with a narrower phylogenetic host population has not conferred significant adaptive advantage for the rhizobia. Possibly the presence of more than one sympatric host species in a given range provides a situation in which the ability to utilize more than one host as symbiont is advantageous to rhizobial survival and increase.

Genetic Base of U.S.-Bred Soybean Cultivars

The development of the present U.S. soybean cultivars in the United States probably has involved two major steps in selection and narrowing of the germplasm base. The first would have involved the domestication of soybeans from the wild predecessor, presumably Glycine soja. The total scope of the selection criteria is, of course, not known, but seed size would appear to have been an im-

portant factor. The second major shift would have occurred during
selection of grain-type plants adapted for mechanical harvest in
the United States. Here selection criteria would have included white
or light colored seed coat, resistance to seed shattering, lodging
resistance, and nonhard seed coat suitable for rapid germination.
This development of the grain-type soybean for North American production resulted in a significant narrowing of the germplasm base.
It is estimated that about 80% of the present soybean germplasm base
in the United States traces to 15 lines introduced from Asia, principally northeast China.

Regardless of the possible historical reasons for the present
status of the rhizobial populations affecting soybean production in
North America, we are now confronted with an indigenous population
that is a mixture of strains of varying efficiency in symbiosis with
our present soybean cultivars.

Strains have been identified that contribute to an important
increase in soybean grain yield over the average strain population
in most fields. If a means could be developed to displace the indigenous rhizobia from symbiosis and establish a precise complemen-
tation of the host crop with desired Rhizobium strains, a significant improvement should be achieved.

GENETIC IMPROVEMENT OF NITROGEN FIXATION

Problem of Indigenous Rhizobia of Lower Efficiency

Let us now direct our attention to the feasibility of enhancing
nitrogen fixation in agronomic production. Data on plant yield indicate that biologically significant interactions occur with specific combinations of host genotypes, with particular Rhizobium
strains in both the self-pollinated pulse crops (soybeans) and cross
pollinated forage legumes. Such interactions may be considered analogous to the combining-ability interactions measured in diallel
analysis of maize hybrids derived from crosses of a series of individual inbred lines. Some R. japonicum strains (such as 110 and
142 from the Beltsville Culture Collection) exhibit what would be
analogous to good "general combining" ability in a diallel. That is,
they form symbioses resulting in superior productivity with an array
of United States-bred soybean cultivars (Abel and Erdman, 1969; Caldwell and Vest, 1970). In addition, while soybean rhizobia have been
subject to natural selection for survival, they have not, until recently, been considered subjects for selective breeding for their
contribution to agronomic yield. Research is now underway directed
toward the breeding of new, highly superior strains of Rhizobium.
However, both in the case of the superior Rhizobium strains now on
hand and the superior strains of Rhizobium that may be produced by
future breeding, a critical problem limits their effective use. This
is the problem of the indigenous Rhizobium strain.

In the case of the soybean host plant, we are able to control
the genotype of crop through our breeding programs and choice of
the cultivar for planting. In the case of the rhizobia, this is
not feasible. Most of the land now used for soybean production in
the United States contains an abundant, heterogeneous population of
Rhizobium strains which will nodulate soybeans. These strains were
probably introduced during the last 80 years or more and, once established in the soil, persist indefinitely whether a soybean crop
is grown or not. Bacteria from this great reservoir of rhizobia
form most of the nodules on the crop plants. Using current inoculation technology, only 5 to 10% of the nodules of a soybean plant
will be occupied with an introduced rhizobium on such land. At
present, no economically feasible method is available to establish
new strains against the competition of the indegenous strains (Ham,
1976; Vest et al., 1973). Therefore, evaluation of Rhizobium
strains in the laboratory and field, studies of host-strain interactions, and genetic engineering programs directed toward breeding
rhizobia for quantum increases in nitrogen fixation will be to no
avail in improving agronomic performance unless a method is found
to establish the desired strains of Rhizobium in symbiosis with the
crop planted.

Given this technological impediment to control of rhizobial
specificity, two general approaches may be taken to dealing with
the genotypes of the Rhizobium populations in the field. We may
accept the indigenous populations as a given factor in the environment and select plant genotypes for adaptation to them, or we may
attempt to control the genotype of the rhizobia admitted to symbiosis and thus take advantage of highly efficient strains.

Manipulating the Host Genotype

In the first case, we may accept the indigenous population as
immutable, in the sense that it cannot be changed or directed by
our efforts (although it certainly is not immutable in the sense
that it is a static population, because we know the population is
dynamic). Under this assumption, we would select crop genotypes
for adaptation to this dynamic microsymbiont population as another
fluctuating environmental variant. However, it would seem that this
is what plant breeders have been doing for many years through the
selection of agronomically superior genotypes which must, in order
to qualify for selection on nitrogen-deficient soils, have necessarily integrated nitrogen fixation with all the other characteristics making up good performance. New techniques for directly
measuring nitrogen fixation per se may improve the efficiency of
selection for this trait.

Selection for physiological traits requires special caution because of possible unanticipated shifts in other characteristics
affecting performance. This is particularly true when the assay

upon which selection is based is performed under artificial conditions or conditions atypical of the production environment. This caution is based not merely upon theoretical considerations but on actual experience. Consider an example from Lotus corniculatus L. (birdsfoot trefoil). Birdsfoot trefoil is a small seeded legume used for pasture and hay. The small seedlings can be difficult to establish in new seedings. Consequently, a breeding program was undertaken to develop seedlings with taller, more vigorous growth habits. The seedlings were evaluated in the greenhouse in a sand growth medium. The tallest seedlings were selected in each generation of selection to be bred as the parents of the next generation in a recurrent selection program of modified mass selection. Seed from each cycle of selection was saved for future study. An evaluation of seed of each cycle, evaluated simultaneously, in a test of seedling vigor, measured under the same conditions as those used in selection (greenhouse sand culture) and using the same criteria as used in selection (seedling height), revealed that selection was effective in increasing plant height. However, when the same cycles were evaluated in conditions simulating field production (field plots), no improvement in stand establishment had been achieved by the selection program that was followed. In fact, the greenhouse selection resulted in a loss of ability to establish a stand. Therefore, whereas selection resulted in modifying the population in the direction of actual selection, it did not achieve the selection goal envisaged. In fact, the opposite occurred, perhaps through the relaxation of selection pressure for adaptation to the multiplicity of stresses associated with survival in the field. It is important to be aware that selection for N_2 fixation conducted exclusively in the laboratory or greenhouse without field selection may also result in loss of field adaptation.

In appraising the opportunity for selection for improved N_2 fixation in the host plant, a fundamental question must be confronted. Is it possible to distinguish the trait? If it is not possible to measure nitrogen fixation, then it can hardly be the subject of selection, unless N_2 fixation is correlated with some other phenotypically distinguishable trait. In my opinion, the question of whether we can distinguish nitrogen fixation is still unresolved. Total plant N can, of course, be determined and, in a N-free growth medium, plant N content should be indicative of N_2 fixed. Such a measure would integrate fixation over the growth period. Other measures, such as acetylene reduction, attempt to measure N_2 fixation at a specific moment in time. Such measures, even if they should be accurate, are representative only of a relatively brief time, which may or may not be reflective of the duration of N_2 fixation. There is evidence of fluctuation in measurements of N_2 fixation in legumes (Trang and Giddens, 1980).

Because of the profound differences among legume crop species –
annuals vs. perennials, diploids vs. polyploids, self-pollinated vs.
cross-pollinated – it is important to caution against the tendency
to attribute information derived from one legume to all legumes.
The amount of N_2 fixed has been reported to vary with stage of
growth and maturation in soybeans (Hardy et al., 1968).   Bergerson
(1970) reported that acetylene-reduction activity exhibited diurnal
variation associated with the length of the light period in soybean
root nodules.   Masterson and Murphy (1976) reported an absence of
diurnal variation in N_2 fixation in white clover (<u>Trifolium repens</u>
L.) when assayed at constant temperature, but a diurnal variation
was clearly evident with a fluctuating ambient day and night temperature differential.   They suggested that the temperature cycle
rather than the photosynthetic cycle in this species is the major
factor responsible for the diurnal variation.   Their field tests
showed that soil temperature was the most important factor affecting
N_2 fixation of white clover in Irish soils.   Gibson and Alston (1981)
reported fluctuation in relative efficiency (RE), defined as 1-(net
H_2 evolution in air/C_2H_2 reduction – H_2 evolution in C_2H_2) in field
grown lupins (<u>Lupinus angustifolius</u>).   The RE declined to a very low
level at the onset of flowering (13 weeks) but subsequently increased
until the last sampling (16 weeks).   It seems that the measurement
of N_2 fixation for any given crop species involves the determination of the most suitable method and timing of measuring and the
most appropriate environmental conditions of measurement, and it is,
therefore, not a simple matter but a complex problem.   The solution
of such problems will require a more detailed descriptive physiology
of the process of N_2 fixation.   It would be prudent to resolve these
problems before expending precious research resources on extensive
selection programs.

Genetic improvement of all biological characteristics, including N_2 fixation, is predicated upon the heritability of the character and the availability of useful genetic variation.   To determine heritability, it is necessary to distinguish that portion of
the expressed variation in a trait resulting from environmental
variation from that portion arising from genetic differences.   To
assess the pertinent environmental variation, it is necessary to
measure fluctuations in a process as they occur in the actual crop
ecosystem.   Large day-to-day fluctuations are known to occur in N_2
fixation.   If diurnal variation exists, the environmental factors
(photoperiod, temperature, etc.) determining that variation need to
be defined.   Descriptive physiology is also needed to define seasonal profiles and correlate N_2 fixation with seasonal growth patterns.   To properly design selection and breeding programs for improved N_2 fixation, it is critical that the environmental factors
affecting N_2 fixation be understood and managed, either through environmental control or experimental design.   It is possible to advantageously manipulate heritability if pertinent environmental
variables are known and controlled.

Present technique for determining N_2 fixation are far from ideal. The acetylene-reduction technique is very sensitive to environmental variation. The ^{15}N techniques (Kohl et al., 1979) are too expensive for plant-breeding programs at traditional funding levels. New techniques are needed for identification and selection of parent lines for crossing and for the evaluation and selection of progeny. The use of plant ureides as a measure of nitrogen fixation has been suggested as a useful technique (LaRue et al., 1981; McNeil, 1981; Patterson and LaRue, 1981; Patterson and LaRue, 1980). Research is needed to develop N_2 fixation assay techniques which are (a) nondestructive of plant, (b) rapid, (c) inexpensive, (d) require a minimum of labor, and (e) applicable in the field.

Controlling Specificity of Microsymbiont Genotypes

We may adopt a second approach to the problem of the indigenous rhizobia and attempt to control the Rhizobium strains admitted to symbiosis. In 1975 at the Fifth American Rhizobium Conference, I proposed a genetic system designed to achieve host-plant specificity of Rhizobium strain symbiosis. This system would require (a) the development of host cultivars which substantially exclude infection by the indigenous Rhizobium strains, (b) the identification or development of Rhizobium strains having the genetic potential to infect these specific host cultivars, and (c) the development of the technology to manipulate the genetic system of Rhizobium in order to couple this specific nodulating ability with high nitrogen fixation (Devine and Weber, 1977). Achievement of these three elements would permit us to supply the crop producers with a production package consisting of cultivars that would not nodulate with the indigenous strains and the inoculum of several superior strains that would nodulate these cultivars.

We have a gene in soybeans, rj_1, the so-called "non-nondulating gene," which, in the homozygous recessive mode, conditions an almost complete inability to nodulate the indigenous Rhizobium in the soil medium (Williams and Lynch, 1954). The rejection of indigenous strains occurs at a very early stage in the nodulation process because no evidence of nodule development is apparent to the naked eye. By backcrossing this gene into our best agronomic lines of soybeans, we can achieve the first requirement - exclusion of the indigenous Rhizobium strains from symbiosis.

We are now attempting to achieve the second requirement. We are searching for genetic information in Rhizobium which will "break down the resistance" of the rj_1 soybean genotype to nodulation and consistently permit a high degree of nodulation under field-production conditions. To obtain strains of Rhizobium with this ability, we plant a five-acre field at Beltsville, Maryland with the "Clark rj_1" soybean genotype. This field has been used for research on soybean rhizobia for many years and contains a large and varied popula-

TABLE 4.   Comparison of the Ability to Nodulate Clark rj_1 Soybeans,
           in Leonard Jar Tests, by <u>Rhizobium</u> Isolates from Nodules
           of Field Grown Clark rj_1 vs. Nodulating Soybeans

		% of isolates					
Isolate source	Eliminated in first screen- ing[a]	Nodules per plant displayed in 2nd screening					
		0	0-1	1-2	2-4	4-8	8+
Clark rj_1	67	5	9	4	7	6	2
Nodulating line	87	6	4	0	2	1	0

[a]Cultures not inducing nodulation in first screening (single Leonard
jar determination) were not tested further. Two Leonard jar determinations were used in the second screening.

tion of rhizobia. After 6 or more weeks of growth, the plants are
mechanically loosened from the soil, one row at a time. The roots
are then removed from the soil and examined for the presence of
nodules. Approximately 200,000 plants have been screened in this
manner, with a yield of about one nodule per 1000 plants in one
year and one per 1500 plants in another year. The nodules recovered
are surface-sterilized and crushed, and the resulting macerated
tissue is used to derive a pure culture of <u>Rhizobium</u> using standard
bacteriological techniques. It is reasoned that an isolate derived
in this manner may have resulted from nodulation by a rhizobial
variant having the genetic capacity to nodulate the rj_1 genotype.
The isolates are then tested for nodulating ability with Clark rj_1
plants in Leonard jars in the greenhouse (Leonard, 1943). Promising
isolates are then tested for performance with Clark rj_1 plants in
soil culture in pot tests. While none of the isolates obtained to
date provide a degree of nodulation of the rj_1 genotype adequate for
economic use, the results are encouraging. Some of the field-derived
strains give a higher level of nodulation than previously tested
strains from the culture collection. A comparison of results obtained from isolates derived from nodules on Clark rj_1 plants with
those from nodules on normally nodulating plants in the same field
indicates that isolates capable of nodulating the rj_1 genotype are
more frequently recovered from the rj_1 plants (Table 4). Therefore,
the Clark rj_1 genotype is an effective screen for such variants
(Devine et al., 1979).

The fact that a highly rj_1 compatible rhizobium has not yet
been found after several years of research suggests that the rj_1
gene is quite stable in expression and should afford dependable exclusion of the indigenous rhizobia when used on a commerical scale.

In another approach to our search for rj_1-compatible rhizobia, we, in cooperation with Dr. Jeffers of the Ohio Agricultural Research and Development Center, are using a field as if it were a large petri dish in a manner analogous to the selection for bacterial mutants with resistance to antibiotics. First, the field was planted to a corn crop to lower the level of soil nitrogen. The following year, Clark rj_1 soybeans were planted. Because of the nitrogen deficiency of the soil and the inability of the Clark rj_1 plants to nodulate, the plant canopy displayed the pale green color typical of nitrogen-deficiency symptoms. Each year, the field is replanted to Clark rj_1 soybeans. If rhizobial variants occur which nodulate rj_1 soybeans, then, presuming that nodulation provides an advantage and favors the proliferation of rhizobia (Kuykendall et al., 1981), the rj_1 compatible rhizobia should increase in number at the site at which they occurred. The following season, other plants in the vicinity should be nodulated, creating the appearance of dark green islands of plants in a field of pale green plants. Nodules from such sites should yield the desired rhizobia.

A variation of the exclusion scheme utilizing incompatibility alleles other than the rj_1 allele has been suggested (Devine and Breithaupt, 1980c). The frequency of the Rj_2 and Rj_4 alleles appears to reflect geographic distribution as discussed earlier. Then, if the Rhizobium strains in a geographic area of the United States are derived, for example, from northern Japan or China and these strains are incompatible with soybean genotypes developed in southeast Asia, it should be possible to avert nodulation with indigenous strains by transferring the genetic control mechanisms for this rhizobial-rejection response to adapted cultivars. In comparing this approach with the use of the rj_1 allele, it should be borne in mind that the rj_1 allele may be readily transferred by backcrossing because it is a single allele, whereas accumulation of several incompatibility alleles, such as Rj_2, Rj_4, etc., is more difficult in a backcrossing program.

Efforts are also underway at the University of Minnesota to employ the system of host-plant exclusion of indigenous strains with selective receptivity to desirable strains-Kvien et al., 1978, 1981). Field screening of a portion of the soybean plant introduction collection has identified lines that do not nodulate normally with the strains indigenous to the Minnesota test site. Some of these soybean lines nodulate well with desirable Rhizobium strains, such as strains 110, 138, and 140. It will be important to determine the completeness of the exclusion of the indigenous strains and the consistency of the exclusion in the soils of the upper Midwest. It would also be important to determine the inheritance of the exclusion mechanisms and the most efficient method of combining this genetic mechanism with improved agricultural crop type for farm production. If the exclusion mechanism is controlled by a large complex of genes located at scattered loci in the genome, the task of

transferring these genes from the agronomically unimproved plant
introductions to agronomically acceptable cultivars will be difficult.

Nitrogen Fixation by Soybeans in Africa

An interesting set of circumstances poses a challenge to increasing N_2 fixation by soybeans in parts of Africa. Soybeans are
not an indigenous crop in Africa, and the United States-bred soybean cultivars do not adequately nodulate and fix nitrogen with indigenous "cowpea-type rhizobia." The introduction of compatible
rhizobia results in markedly improved fixation. However, lack of
skilled personnel and laboratories in many developing nations re-
stricts the production, distribution, and use of inoculum. Nangju
(1980) suggested a promising solution to this problem. He found
that soybeans from southeast Asia nodulated successfully with indigenous African rhizobia and the resulting symbiosis was effective
in N_2 fixation. The United States-bred cultivars, however, have the
advantage of higher grain yield potential and better resistance to
lodging and seed shattering. Nangju (1980) suggested breeding to
combine the desirable agronomic characteristics of the United States-
bred cultivars with the symbiotic potential for N_2 fixation with the
indigenous rhizobia characteristic of the southeast Asian germplasm.

Characteristic of the rj_1 Allele

Because of the potential utility of the rj_1 gene, we are interested in learning more about its function. It is important to know
whether the rj_1 gene conditions a general antagonism to rhizobial
growth and metabolism or has only the more limited effect of re-
stricting nodulation. Clark (1957) reported equal numbers of rhizobia were recovered from the roots of rj_1 and Rj_1 plants. Elkan
(1962) found a higher number of rhizobia in the rhizosphere of the
rj_1 plants during most of the 60-day growing period sampled. Devine
and Weber (1977) found that some Rhizobium strains nodulated rj_1
plants and were effective in fixing nitrogen once nodulation was
achieved. Thus, the incompatibility conditioned by the rj_1 gene
does not appear to be a general antagonism to rhizobial metabolism
and function but appears to be specific to some early process in
the infection event per se. It is reasonable for us to expect that
once nodulation has been achieved, N_2 fixation will not be impaired.

The degree of nodulation achieved by the rj_1-compatible strains
found to date is not deemed adequate for commercial use. The several Rhizobium strains known to produce a limited degree of nodulation of rj_1 plants in sand or vermiculite culture produce a much
more limited degree of nodulation in soil culture (Clark, 1957;
Devine and Weber, 1977). It is not possible to predict the characteristics of the highly rj_1 compatible strains we are seeking on
the basis of the characteristics of the presently available strains

TABLE 5. Effect of the Rhizobitoxine Analog, Aminoethoxyvinylglycine (AVG), on Nodulation of Clark and Clark rj₁rj₁ Soybeans

AVG Concentration (M)	Clark rj₁rj₁				Clark			
	Strain 61		Strain 110		Strain 61		Strain 110	
	No. of plants	Nodules/plant	No. of plants	Nodules/plant	No. of plants	Nodules/plant	No. of plants	Nodules/plant
0	37	7.4	38	0	3	17.3	4	16.0
$5 \cdot 10^{-10}$	30	5.8	35	0	4	14.0	3	15.0
$5 \cdot 10^{-8}$	33	6.8	37	0	4	12.3	3	10.3
$5 \cdot 10^{-6}$	36	5.1	36	0	3	13.3	4	10.0
$5 \cdot 10^{-5}$	36	6.3	39	0	4	19.3	4	13.3
$5 \cdot 10^{-4}$	35	0.8	38	0	2	0.7	–	–

TABLE 6.   Comparison of the Effect of Temperature on the Nodulation
of rj_1 Soybeans and Rhizobial Induced Foliar Chlorosis

Property measured	Rhizobium strain	21°C	27°C	32°C
Nodules per plant[a]	61	7.5 a*	6.4 a	3.1 bc
	76	2.2 cd	3.8 b	1.6 d
Rhizobial-induced foliar chlorosis[b]	61	1.0 e*	2.8 abc	2.0 cde
	76	1.2 e	3.6 a	2.6 abc

*Means followed by different letters are significantly difference
at P ≤ 0.05.
[a]Clark rj_1 soybeans nodulated by rj_1-compatible rhizobia.
[b]Scored 1 to 5 (1, no symptoms; 5, most severe chlorosis).   Hawkeye
soybeans.

with limited rj_1 compatibility.   However, we have observed some interesting relationships using these strains.   Some strains of Rhizobium produce a chemical, termed rhizobitoxine, that causes a foliar
chlorosis in soybean plants (Johnson and Clark, 1958; Johnson et al.,
1958; Owens et al., 1972; Owens and Wright, 1965).   The Rhizobium strains
capable of nodulating the rj_1 genotype also have a propensity to
produce symptoms of rhizobitoxine-induced chlorosis in soybean foli-
age (Devine and Weber, 1977).   This associated occurrence of the two
characteristics prompted us to test an ethoxy analog of rhizobitoxine (aminoethoxyvinylglycine), which was available in sufficient
quantity, for possible modification of the ability of rhizobia to
nodulate the Clark rj_1 genotype of soybeans.   The analog was known
to be similar to rhizobitoxine in the ability to induce foliar chlorosis and inhibit ethylene production.   Exogenous addition of the
analog (at concentrations ranging from 0 to 5×10^{-4} M) to cultures
of a Rhizobium strain (strain 110) not capable of infecting rj_1 soybeans did not endow the cultures with infectivity (Table 5) (Devine
and Breithaupt, 1980a).   Addition of the analog to cultures of the
rj_1-compatible strain 61 did not enhance the nodulating potential of
this strain.

Temperature in the range 21 to 32°C was observed to alter the
phenotypic nodulation response of the rj_1 gene with rj_1-compatible
rhizobia, strains 61 and 76 of the Beltsville Culture Collection
(Devine and Breithaupt, 1980b).   There was no significant difference
in the number of nodules per plant formed by Rhizobium strain 61 on
Clark rj_1 plants cultured at 21 vs. 27°C (Table 6).   At 32°C, nodulation by strain 61 was significantly reduced.   For Rhizobium strain
76, the highest degree of nodulation occurred at 27°C, with nodula-

tion at 21 and 32°C significantly lower. If the effect of temperature was exclusively a direct effect on the Clark rj_1 plants, then
the nodulation induced by both strain 61 and 76 would have shown
the same pattern of reaction to temperature. This did not occur.
With strain 61, there were no significant differences in degree of
nodulation at 21 and 27°C. In contrast, with strain 76, nodulation
was significantly lower at 21°C than at 27°C. The cooler temperature reduced the nodulating potential of strain 76 but not strain
61. At the highest temperature, 32°C, nodulation was low for both
strains, and this may have been a plant response. The induction of
symptoms of foliar chlorosis by the same two Rhizobium strains was
either nonexistent or lowest at 21°C (Table 6). Chlorosis was quite
apparent at 27 and 32°C. Thus, although the rhizobitoxine-induced
chlorosis symptoms were quite marked at 32°C, nodulation was lowest
at this temperature. The differential effect of temperature on
these two processes suggested that, whereas rhizobitoxine is correlated with both processes, separate control mechanisms with different temperature optima govern their expression. This suggests
that the induction of rhizobitoxine symptoms and ability to nodulate the rj_1 genotype are not conditioned as a pleiotropic effect of
a single rhizobial gene. Indeed, the variation in the intensity of
expression of rhizobitoxine symptoms with different Rhizobium strains
on the same host genotype suggests either multiple loci or a multiple
allelic series controlling the trait in the rhizobia. It may indeed
be that the association of these two characteristics in rhizobia is
not the result of an intrinsic physiological relationship, but of
the fixing of these two characteristics in the same populations by
"random drift" during the course of evolution. The strains infective on rj_1 may have the potential for infection by paths alternative to the root hair infection, perhaps by wounds. However, once
the bacteria have gained entrance, the metabolic integration of
the two genetic complements may be defective, and rhizobitoxine
build up may be a symptom of this defective homology.

We are also interested in the ability of the rj_1 genotype to
distinguish between the presently available rj_1-compatible strains
and the rj_1-incompatible strains in nodulation response. The rj_1-
compatible strain 61 was marked with genetic resistance to nalidixic
acid by mutant selection, and the rj_1-incompatible strain was marked
with resistance to rifampicin and streptomycin. Broth cultures of
the two strains were mixed before inoculation of the rj_1 plants
(Devine et al., 1980). The nodules produced on the rj_1 plants were
then examined for the presence of the marker genes. Fifty-five
nodules produced cultures that would grow exclusively on a nalidixic
acid-containing medium, 61 produced cultures that would grow only on
a medium with rifampicin and streptomycin, and 54 nodules produced
cultures that would grow on both media. A subsequent test of the
cultures derived from the nodule isolates which grew on the medium
with rifampicin and streptomycin for the ability to nodulate rj_1
plants in pure culture indicated that these strains had not acquired

a permanent modification enabling them to nodulate the rj_1 genotype. Apparently, they were able to nodulate the rj_1 plants only because of the transitory association with the rj_1-compatible strain 61 Nal^R. If the nodules yielding cultures that grew on both the media had resulted from rhizobia that were the products of genetic recombination of these alleles, they should also yield cultures that would grow on a medium containing all three antibiotics. A sample of 10 such nodule isolates were tested on media with the three antibiotics, and none grew. We concluded that this class of nodules contained a mixture of two genetically distinct strains, 61 Nal^R and I-110 ARS.

The fact that 36% of the nodules formed after inoculation with the mixed inoculum apparently contained only the strain I-110 ARS, which does not nodulate rj_1 plants when used in pure culture, suggests that either (a) some diffusible product of strain 61 Nal^R was absorbed by cells of strain I-110 ARS and transiently endowed them with infectivity or (b) strain 61 Nal^R acted on the root cells and rendered them receptive to infection by any Rhizobium in contact with an infection site.

In a subsequent series of experiments, we found that an rj_1-incompatible strain remained unchanged in ability to nodulate rj_1 plants when (a) grown in culture filtrates of rj_1-compatible strains, (b) cultured in U-tubes across a 0.08-mm pore size polycarbonate membrane from rj_1-compatible strains, or (c) cultured in dialysis tubing suspended in broth cultures of an rj_1-compatible strain (Devine et al., 1981). No evidence was found of a diffusible compound produced in vitro that was capable of altering the ability of rj_1-incompatible strains to nodulate rj_1 plants. Therefore, it appears that direct contact with the nodulation site is required for rj_1-compatible strains to facilitate nodulation by rj_1-incompatible strains.

CONCLUSIONS AND OPPORTUNITIES

The improved agronomic performance achieved in soybean production with the use of some Rhizobium strains indicates the potential for increasing crop performance with better adapted, more compatible rhizobia. It should be emphasized, however, that the identification or development of agronomically superior rhizobia will be to no avail in improving crop production unless means can be developed for establishing the desired Rhizobium strains in symbiosis with the host crop. It is critical that such means be found, and research in this area is urgently needed. The use of host cultivars that exclude the indigenous rhizobia while admitting selected strains is a promising and inexpensive method to achieve this objective. The exclusion of the indigenous strains provided by the rj_1 gene is extremely powerful. It is possible that a lesser degree of exclusion by the host

It is possible that a lesser degree of exclusion by the host would prove adequate for practical use. Possibly there would be a greater probability of finding a <u>Rhizobium</u> strain capable of nodulating a host genotype with a lesser degree of exclusion than with the rj_1 allele as now expressed in Clark rj_1. In a promising development in this area, we have been successful in obtaining lines of soybeans, derived from hybridization of the rj_1 genotype with another genetic stock, that exhibit a modification of the expression of the rj_1 allele (Devine, unpublished). Hopefully, continued research will provide a functional system for precise determination of the genotypes of host and microsymbiont involved in symbiosis.

REFERENCES

G. H. Able and L. W. Erdman, Agron. J., 56:423-424 (1964).

F. J. Bergersen, Aust. J. Biol. Sci., 23:1015-1025 (1970).

J. E. Beringer, N. J. Brewin, and A. W. B. Johnston, Heredity, 45: 161-186 (1980).

B. E. Caldwell, Crop Sci., 6:427-428 (1966).

B. E. Caldwell and G. Vest, Crop Sci., 10:19-21 (1970).

B. E. Caldwell and G. Vest, Crop Sci., 8:680-682 (1968).

F. E. Clark, Can. J. Microbiol., 3:113-123 (1957).

T. E. Devine, in: Genetic Engineering for Nitrogen Fixation (A. H. Hollaender et al., ed.), Plenum Publishing Corp., New York, pp. 417-418 (1977).

T. E. Devine, Agron. Abstr., p. 160 (1981).

T. E. Devine and B. H. Breithaupt, Crop Sci., 20:819-821 (1980a).

T. E. Devine and B. H. Breithaupt, Crop Sci., 20:394-396 (1980b).

T. E. Devine and B. H. Breithaupt, Crop Sci., 20:269-271 (1980c).

T. E. Devine and B. H. Breithaupt, USDA Tech. Bull., no. 1628 (1981).

T. E. Devine, B. H. Breithaupt, and L. D. Kuykendall, Crop Sci., 21: 696-699 (1981).

T. E. Devine, L. D. Kuykendall, and B. H. Breithaupt, Agron. Abstr., p. 60 (1979).

T. E. Devine, L. D. Kuykendall, and B. H. Breithaupt, Can. J. Microbiol., 26:179-182 (1980).

T. E. Devine and W. W. Reisinger, Agron. J., 70:510-511 (1978).

T. E. Devine and D. F. Weber, Euphytica, 26:527-535 (1977).

R. O. D. Dixon, Arch. Mikrobiol., 85:193-201 (1972).

G. H. Elkan, Can. J. Microbiol., 8:79-87 (1962).

H. J. Evans, J. E. Lepo, F. J. Hanus, K. Purchit, and S. A. Russell, in: Genetic Engineering of Symbiotic Nitrogen Fixation and Conservation of Fixed Nitrogen (J. M. Lyons, R. C. Valentine, D. A. Phillips, D. W. Rains, and R. C. Huffaker, eds.), Plenum Press, New York (1981).

A. H. Gibson, B. L. Dreyfus, R. J. Lawn, J. I. Sprent, and G. L. Turner, in: Current Perspectives in Nitrogen Fixation (A. H. Gibson and W. E. Newton, eds.), Aust. Acad. Sci., Canberra, p. 373 (1981).

P. R. Gibson and A. M. Alston, in: Current Perspectives in Nitro-
     gen Fixation (A. H. Gibson and W. E. Newton, eds.), Aust. Acad.
     Sci., Canberra, p. 375 (1981).

G. E. Ham, in: World Soybean Research (L. D. Hill, ed.), Inter-
     state Printers and Publishers, Danville, Ill., p. 144-150
     (1976).

G. E. Ham, L. R. Frederick, and I. C. Anderson, Agron. J., 63:69-72
     (1971).

F. J. Hanus, S. L. Albrecht, R. M. Zablotowicz, D. W. Emerich, S. A.
     Russell, and H. J. Evans, Agron. J., 73:368-372 (1981).

R. W. F. Hardy, R. D. Holsten, E. K. Jackson, and R. C. Burns, Plant
     Physiol., 43:1185-1207 (1968).

H. W. Johnson and F. E. Clark, Soil Sci. Soc. Am. Proc., 22:527-528
     (1958).

H. W. Johnson and U. M. Means, Agron. J., 52:651-654 (1960).

H. W. Johnson, U. M. Means, and F. E. Clark, Agron. J., 50:571-574
     (1958).

H. H. Keyser, Biological Nitrogen Fixation Technology for Tropical
     Agriculture, Cali, Colombia, p. 44 (1981).

H. H. Keyser, B. B. Bohlool, T. S. Hu, and D. F. Weber, Proc. 8th
     North Am. Rhiz. Conf., p. 72 (1981a).

H. H. Keyser, P. van Berkum, and D. F. Weber, Agron. Abst., p. 163
     (1981b).

H. H. Keyser, P. van Berkum, and D. Weber, in: Current Perspec-
     tives in Nitrogen Fixation (A. H. Gibson and W. E. Newton,
     eds.), Aust. Acad. Sci., Canberra, p. 374 (1981c).

D. H. Kohl, G. B. Shearer, and J. E. Harper, Agron. Abst., p. 159
     (1979).

L. D. Kuykendall, T. E. Devine, and P. B. Cregan, Abst. Annu. Meet-
     ing, Am. Soc. Microbiol., Abst. I20 (1981).

C. Kvien, G. E. Ham, and J. W. Lambert, Agron. Abst., p. 142 (1978).

C. S. Kvien, G. E. Ham, and J. W. Lambert, Agron. J., 73:900-905
     (1981).

LaRue, T. T. Patterson, and R. Glenister, Plant Physiol. (Abst.),
     67:78 (1981).

L. T. Leonard, Soil Sci., 15:277-283 (1923).

L. T. Leonard, J. Bacteriol., 45:523-527 (1943).

S. T. Lim, S. L. Uratsu, D. F. Weber, and H. H. Keyser, Genetic
     Engineering of Symbiotic Nitrogen Fixation and Conservation
     of Fixed Nitrogen (J. M. Lyons, R. C. Valentine, D. A. Phillips,
     D. W. Rains, and R. C. Huffaker, eds.), Plenum Press, New York
     (1981).

C. L. Masterson and P. M. Murphy, in: Symbiotic Nitrogen Fixation
     in Plants (P. S. Nutman, ed.), Cambridge Univ. Press, London
     (1976).

C. L. Masterson and M. T. Sherwood, Irish J. Agric. Res., 13:91-99
     (1974).

D. L. McNeil, Biological Nitrogen Fixation Technology for Tropical
     Agriculture, Cali, Columbia, p. 48 (1981).

F. Munevar and A. G. Wollum, II, Agron. Abst., p. 158 (1980).

D. Nangju, Agron. J., 72:403-406 (1980).

L. D. Owens, J. F. Thompson, R. G. Pitcher, and T. Williams, J. C.
    S. Chem. Comm., 1972:714 (1972).

L. D. Owens and D. A. Wright, Plant Physiol., 40:931-933 (1965).

T. G. Patterson and T. A. LaRue, Agron. Abst., p. 90 (1980).

T. Patterson and T. LaRue, Plant Physiol. (Abst.), 67:78 (1981).

K. R. Schubert and H. J. Evans, Proc. Nat. Acad. Sci. (U.S.), 73:
    1207-1211 (1976).

K. M. Trang and J. Giddens, Agron. J., 72:305-308 (1980).

G. Vest, Crop Sci., 10:34-35 (1970).

G. Vest and B. E. Caldwell, Crop Sci., 12:692-693 (1972).

G. Vest, D. F. Weber, and C. Sloger, in: Soybeans: Improvement,
    Production, and Uses (B. E. Caldwell, ed.), Am. Soc. of Agron.,
    Madison, Wisconsin (1973).

L. F. Williams and D. L. Lynch, Agron. J., 46:28-29 (1954).

PRINCIPLES OF Rhizobium STRAIN SELECTION

Jake Halliday

University of Hawaii NifTAL Project
P. O. Box O
Paia, Maui, Hawaii 96779. U.S.A.

INTRODUCTION

The following account outlines one approach to the selection of Rhizobium strains for use in legume seed inoculants. The procedures described were used successfully in a specific program concerned with the selection of appropriate rhizobia for forage legume introductions in acid, infertile soils of tropical Latin America (Halliday, 1979). The principles underlying the approach apply equally well to other programs, and some examples of alternative methodologies are mentioned in the text. Provided they take account of the underlying principles of Rhizobium strain selection stressed in this article, individual investigators can modify the techniques and improvise with equipment to suit their own purposes and the facilities available to them.

OBJECTIVE OF STRAIN SELECTION

Strain selection is performed to ensure that a legume seed inoculant contains a strain, or strains, of Rhizobium capable of forming fully effective, N_2-fixing nodules on the legume species for which it is recommended and under the conditions of soil and climate in which the legume crop is grown.

CHARACTERISTICS OF IDEOTYPIC STRAINS OF Rhizobium

Some characteristics of strains of Rhizobium to be used as legume inoculants can be regarded as "essential," whereas others are "desirable" depending on the specific selection objective.

155

One essential characteristic is the ability to nodulate the legume crop of interest in the field conditions under which it is grown. Such strains are referred to as infective. Strains of Rhizobium which are infective in the field will usually have exhibited competitive ability if they displaced nodulation by native strains present at the site. They will also have been stress tolerant if they successfully nodulated legumes in soils with excesses or deficiencies in their physical and chemical composition.

A second essential characteristic is that the strain be able to fix sufficient N_2 to sustain a level of legume production close to, or surpassing, the production possible if the legume were supplied with nitrogenous fertilizers. Such strains are referred to as effective. Strains which are fully effective are usually carbon efficient and hydrogen efficient as well. The "efficiency" of a Rhizobium is seldom measured during strain selection, and use of the term should be avoided. Effectiveness is usually what is meant.

A third essential character of an ideotypic Rhizobium strain is that it should perform satisfactorily when subjected to the component processes of commercial-scale inoculant-production systems. Inoculant strains must multiply well in bulk culture and be able to mature to high populations in the carrier material, which is usually peat.

A fourth essential character is ability to survive well during distribution to, and use by, farmers. Strains should be tolerant to the anticipated maximum temperature that they will encounter. They must also survive well during the seed or soil inoculation procedures used by farmers. Additionally, they must survive on seed in soil from the time of their application until the emerging legume radicle is susceptible to infection (usually at least 7 days).

Characteristics which are in the "desirable" category are long term persistence and fungicide and insecticide tolerance.

Long term persistence is expected of strains of Rhizobium used to inoculate perennial forages. Implicit in the concept of persistence is saprophytic competence, a summary term for all those traits that permit a Rhizobium strain to live as a stable member of the soil microflora, even in the absence of its legume host. Persistence of strains for annual crop legumes from season to season may be considered a desirable trait in some circumstances as it obviates the need for inoculation in subsequent years. But there may be cropping systems in which carry-over strains from a previous crop may nodulate a following crop relatively ineffectively and even outcompete effective introduced strains. This can occur in rotations of soybean with legumes such as peanut and cowpea that nodulate with the cowpea miscellany of rhizobia.

Fungicide or insecticide resistance may be a desirable trait when the normal practice is to sow legume seeds pretreated with these substances, some of which are toxic to most strains of Rhizobium.

SCIENTIFIC BASIS FOR STRAIN SELECTION

Rhizobium strains vary widely in the characteristics listed above. Some strains nodulate some genera, or species, or varieties of legumes and not others. This has given rise to the durable, but highly criticized, taxonomy of rhizobia based on their cross-inoculation affinities. Among the strains capable of infecting and nodulating a particular legume, there is great variation in the amount of N_2 they fix; i.e., variation in effectiveness. There is considerable strain variation in the other listed traits as well, and thus an opportunity exists to select superior strains. Unlike higher plants which can be improved through breeding and hybridization, Rhizobium improvement is currently practical only by selection from natural populations.

WHEN IS STRAIN SELECTION JUSTIFIED?

As will be appreciated from the following procedures, the selection of superior Rhizobium strains is a lengthy undertaking. Several years of study may be necessary to complete characterization and testing. Given that strains of Rhizobium for many legumes have already been developed in research laboratories around the world, it makes sense to obtain and use these, rather than to initiate an extensive selection program. Selection of rhizobia is only really justified when the specific selection objective cannot be satisfied by strains held in existing collections. Examples of circumstances under which strain selection may be required are as follows:

a) When the legume of interest is an uncommon species for which there is no recommended inoculant strain.

b) When inoculation of the particular legume with recommended strains of Rhizobium under field conditions fails to give adequate nodulation and N_2 fixation. This can occur if the legume variety is different from that with which the inoculant strain was developed or if the soil and climatic conditions vary from those under which the inoculant was developed.

SELECTION FOR SOIL STRESS TOLERANCE

This paper describes a step-wise selection procedure for the development of a <u>Rhizobium</u> strain recommendation for legumes planted under a particular soil condition. This approach is unconventional in the sense that strains of <u>Rhizobium</u> in current use as legume seed inoculants are developed for the species of legume with which they will be used, rather than the soil type in which the legume will be grown. In the technologically advanced countries, it is normal farm practice to modify soil conditions to make them suitable for a particular crop. It is not unreasonable, therefore, to expect a rhizobial inoculant for a legume species to perform well wherever that legume is grown. In the developing nations, however, soil amendment is minimal or not practiced at all, and crop plants are often grown under stresses of adverse soil factors that cannot be economically alleviated. It may be unreasonable to expect that a single strain of <u>Rhizobium</u> will perform equally well as an inoculant in the wide array of soil types under which its host legume is grown in the tropics. One reason that legume inoculation is not widely successful in developing countries is that available inoculants obtained from the United States, Australia, or elsewhere do not have strains selected for and adapted to, the extremes of soil stress encountered in the tropics.

There is a widely held view that strain selection and legume inoculation have little potential for improving yields of tropical legumes since tropical legumes are not specific in their <u>Rhizobium</u> strain requirements, and because suitable rhizobia occur universally in tropical soils. There are a few notable exceptions, such as soybeans and leucaena, and thus two categories of tropical legumes were recognized. The promiscuous (P) group can be nodulated by a wide array of strains of tropical rhizobia. The specific (S) group requires specific rhizobial strains for nodulation. The majority of tropical legumes were judged to belong to the P group, and it has been generalized that it is unnecessary to inoculate these legumes with rhizobia, as no benefit would be expected.

The grouping of tropical legumes simply as S or P types is no longer tenable nor useful. Many tropical legumes previously placed in the P group are now known to form fully effective (i.e., high N_2-fixing) symbioses with only a few strains out of the diverse array of rhizobia that can nodulate them. Thus a distinction is drawn between this promiscuous-ineffective (PI) group of legumes and the promiscuous-effective (PE) group (Date and Halliday, 1980). Studies of the <u>Rhizobium</u> affinities of tropical forage legumes, for example, reveal that a majority of them are in the PI group, suggesting a potential for increasing their production by providing appropriate strains of <u>Rhizobium</u>.

The important role played by stress factors in tropical soils as modifiers of symbiotic performance is now recognized. Thus, tropical legumes can and do benefit from inoculation when strains are selected specifically for the particular variety of legume being planted and for tolerance of the soil conditions in which that legume is to be grown.

DEFINITION OF THE SPECIFIC SELECTION OBJECTIVE

No strain selection program should be undertaken without clear definition of the specific selection objective(s). The methods of selection employed may need to be modified to suit the objective. The specific selection objective for which the procedures that follow were developed was to select strains of Rhizobium able to nodulate and fix N_2 in association with acid-tolerant legume accessions being introduced to the acid, infertile soils of Latin America.

THE MULTI-STAGE SCREENING APPROACH
TO Rhizobium SELECTION

Successful selection of superior rhizobia is favored if the number of strains from which the selection is made is large and diverse. The most meaningful test of Rhizobium performance is field evaluation since this is an integrated appraisal of the various traits that make up a successful inoculant strain. However, the management of field trials to select rhizobia is difficult and costly, even when the number of strains under test is small. Adopting a multi-stage screening procedure that progressively eliminates undesirable strains from an initially high number of contenders to a relatively small number of promising strains for testing at the field level, is one way to reconcile the requirements that selection be from a diverse genetic base, and that strains also be assessed under field conditions.

PRESELECTION OF STRAINS ENTERING
THE SCREENING PROCEDURE

It is advisable to include in the screening procedure strains of Rhizobium that originated from a diverse array of host-plant germplasm and that are representative of diverse geographic regions. However, some reduction in the number of strains can be made based on what is known from other selection programs. In general, rhizobia isolated originally from the same genus, and sometimes species, as the legume for which a superior strain is being sought emerge from selection programs as the best strains for use in legume inoculants. Also, when the specific selection objective includes tolerance to a particular soil stress or climatic condition, rhizobia isolated from

legumes growing under those conditions are the most likely to be
rated highly in the selection process. Hopefully, there is a Rhizobium collection or collections of authenticated strains of known
origin available to the investigator. Otherwise, a collection of
strains has to be assembled. Only after checking whether likely
strains are available from existing Rhizobium collections, such as
the Rhizobium Germplasm Resource at NifTAL, should collection and
isolation of new strains be contemplated. Detailed procedures for
the collection, isolation, purification, authentication, characterization, and preservation of strains of Rhizobium are described elsewhere (Date and Halliday, 1979; Somasegaran et al., 1979). Preselec-
tion of strains with suitable backgrounds should aim to generate a
cluster of 50 to 100 rhizobia that will feed into stage I of the
strain-selection procedure.

Stage I: Screening for Genetic Compatibility

In this stage, strains of Rhizobium are screened for ability to
nodulate the legume of interest. The test used involves a high degree of bacteriological control and is suited to handling large numbers of strains. The system most commonly used is based on growth
tubes in which seedlings are raised in a solid nutrient medium under
artificial illumination. Seeds must be surface sterilized, usually
with concentrated sulfuric acid, hypochlorite, or acidified mercuric
chloride. They are pregerminated in inverted, sterile petri dishes
containing water agar. When the radicles are 3 to 5 mm long, uniform seedlings are transferred aseptically to tubes containing agar
deeps (or slants). Tubes are routinely 2.5 × 25 cm and are capped
with a plug of muslin-wrapped cotton wool. Aliquots of 1 ml of suspension of the test strains are added to each tube either at trans-
planting or 3 to 5 days later. At least three replications of each
strain treatment are essential, and five are preferred. Roots of
seedlings should be shielded from light and this is best accomplished
by using the arrangement in Fig. 1. Alternatively, tubes may be
wrapped in Al foil. Two control treatments are required. In one
case, the plants are "inoculated" with sterile water only (uninoculated control) and in the other case, they are provided with 70 ppm
of N as an ammonium nitrate (or potassium nitrate) solution (plus
nitrogen control). Tubes are scored at intervals for the presence
or absence of nodules. A word of caution: with many tropical
legumes, it is common to observe tumor- or callus-like outgrowths
on roots of seedlings raised in growth tubes. These outgrowths occur in the presence or absence of rhizobia and are not nodules. They
cannot usually be distinguished from nodules by eye, and it is
strongly recommended that plants are harvested from tubes and checked
under a binocular microscope for real nodules. "Apparent" nodules
lack structural organization and leghemoglobin. Timing of the harvest varies, depending on legume species, but the harvest will usually
be about 35 days after sowing.

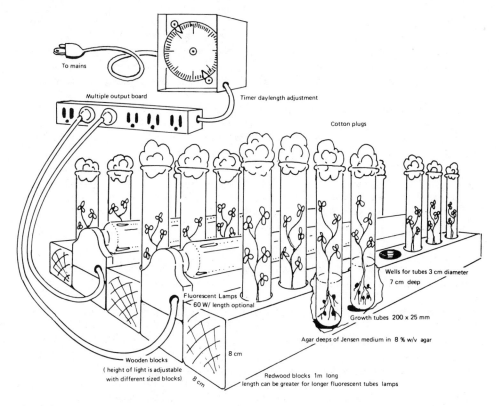

To mains

Multiple output board

Timer daylength adjustment

Cotton plugs

Wells for tubes 3 cm diameter
7 cm deep

Fluorescent Lamps
60 W/ length optional

Growth tubes 200 x 25 mm

Agar deeps of Jensen medium in 8 % w/v agar

8 cm

Wooden blocks
(height of light is adjustable
with different sized blocks)

8 cm

Redwood blocks 1m long
length can be greater for longer fluorescent tubes lamps

Fig. 1. Low-cost light rack for culturing plants used in plant infection tests.

Some investigators place significance on other data taken on
plants grown in growth tubes. "Earliness to nodulation" may be of
some value. It is inappropriate, however, to attribute relative N_2
fixation effectiveness to strains based on N accumulation in plants
raised under such artificial conditions. The root medium and atmospheric composition within plugged test tubes differ from those
which the plants require for optimum performance and may constrain
expression of N_2-fixing potential.

Alternate methodologies are required for large-seeded species
that quickly become cramped in growth tubes. These include the use
of growth pouches or "Gibson" tubes. Growth pouches are made of
autoclavable plastic and have an absorbent towel insert. Seedlings
germinate in a fold (or are pregerminated and transplanted into the
fold) at the upper rim of the pouch. Roots develop within the
pouch, where they are nourished by a nutrient medium, and plant tops
grow in the open air. The method offers the advantage that effective

nodulation can be reliably determined, but caution in attributing
relative effectiveness of strains on the basis of a pouch test is
necessary. Modifications of the method include subdividing the
pouches with heat bonding to permit a single pouch to be used for
several strain treatments, or replications of the same treatment.

The "Gibson" tube contains a long agar slant that reaches to
the upper rim of the tube, and the tube is filled to the rim with
liquid medium or sterile water. They are capped with Al foil.
Radicles of pregerminated seedlings are introduced through a small
orifice in the Al. The roots develop inside the tube, and the plant
tops grow outside the tube. The method offers similar advantages to
those of pouches, namely that effective nodulation shows up readily.
Modifications of "Gibson" tubes include omission of the liquid
phase or half filling the tubes.

Obviously nodulation in the uninoculated control treatments in
stage 1 raises concern about inadequate bacteriological control and
invalidates the experiment.

Some texts advocate dedication of entire light rooms for the
culture of plants in growth tubes. Most workers will find a low-
cost system, such as that in Fig. 1, more than adequate for their
needs. The system is highly flexible and can be readily modified
to serve for pouches or "Gibson" tubes that require overhead illumination. The issue of light quality has been overplayed, and researchers should not be preoccupied with obtaining special spectrum
lamps or with including incandescent lamps in the system. Regular
domestic fluorescent lamps have served satisfactorily in the screening procedure described here.

Stage II: Screening for N_2-Fixation Effectiveness

In this stage, the objective is to rank infective strains from
stage I in order of potential N_2-fixation effectiveness with the
legume species or cultivar of interest. Theoretically, in this test
there should be no factors limiting growth of the legume except nitrogen, so that full expression of each strain's N_2-fixation effectiveness is possible. In practice, it is assumed that the nutrient
regime and other aspects of growth conditions are not limiting, even
though there are known examples of legumes for which standard conditions are not nonlimiting. Sand-jar assembles are used in this
test because they permit more realistic growth conditions than tubes,
pouches, etc., but they retain the high degree of bacteriological
control which is still essential if valid results are to be expected.

The Leonard jar is one example of such a sand-jar assembly and
is depicted in Fig. 2. Watering is the most common source of contamination in Rhizobium strain testing in pots and in the field.

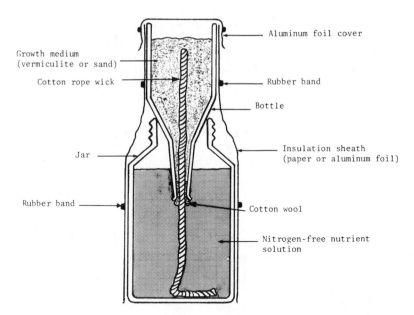

- Aluminum foil cover
- Growth medium (vermiculite or sand)
- Cotton rope wick
- Rubber band
- Bottle
- Insulation sheath (paper or aluminum foil)
- Jar
- Rubber band
- Cotton wool
- Nitrogen-free nutrient solution

Fig. 2.   One example of a Leonard jar assembly. Provided the basic design is followed, the system can be modified to permit use of cheap, locally available materials. The entire assemblies with nutrient solution are autoclaved at 121°C for 1-2 hours, depending on their size. Anticipate some losses (5%) due to breakage of glass components on heating and cooling.

Leonard-type sand jars greatly reduce the frequency of watering and are, therefore, less prone to contamination. Sand jars are easily constructed from locally available materials, but they have the disadvantage that sterilizing them requires a very large autoclave.

As with growth tubes, surface sterilized pregerminated seeds are sown in the sand jars. Four seedlings are allowed to establish and are thinned later to two by snipping off the tops. Drops (standardized rate) of suspensions of strains of Rhizobium are added in the jars 5 days after sowing (one strain per jar). Plants are harvested destructively at a time after sowing that depends on the legume species under test. Usually 60 days after sowing is appropriate.

Date taken on sand jar experiments vary from investigation to investigation and include the following: nodule number, nodule dry weight and/or fresh weight, nodule color, nodule distribution, total plant fresh or dry weight, top weight (fresh or dry), root weight (fresh or dry), acetylene-reduction rate, percentage N in tissues, and total N produced. Of these, total N produced is the most mean-

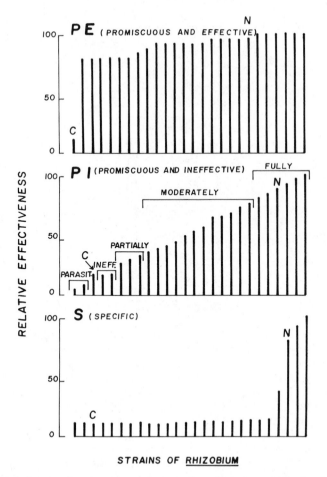

Fig. 3.   Hypothetical ranking of strains for potential nitrogen-
          fixing effectiveness as determined in sand jar assemblies
          (stage II).

ingful integration of N_2-fixation effectiveness over time, and as
this is highly correlated with total plant dry weight, a reliable
measure of relative effectiveness of strains of Rhizobium is possible with nothing more sophisticated nor costly than a common balance.

The main problem encountered with this test relates to overheating in the greenhouses or growth rooms where the experiments are
performed. Most of the sand-jar trials observed by the author in
the tropics are, in fact, selecting high temperature-tolerant rhizobia at the same time! Other problems relate to the occasional failure of the irrigation from beneath, which depends on capillary rise,
and breakage of glass components in autoclaving and handling.

A ranking of strains based on hypothetical yields from stage
II trials is presented in Fig. 3. The demarcation of effectiveness
categories is somewhat subjective but is nevertheless useful.
Strains are assessed relative to the uninoculated control and the
N_2 control and described as (in ascending order of merit) parasitic,
ineffective, partially effective, moderately effective, or fully
effective.

Ordinarily about 30 to 50 strains would be evaluated at stage
II in Leonard jars. Three replications are essential, and five are
preferred. The top ten strains are chosen for further screening at
stage III.

The principal merit of Leonard jar trials are that data on the
potential effectiveness of strains of <u>Rhizobium</u> with a particular
legume will be upheld in independent screening trials by other investigators. Thus, researchers can exchange information that is
stable and demonstrable on the N_2-fixing potential of strains. Pot
and field tests, on the other hand, give information of the plant-
<u>Rhizobium</u>-soil interaction that may or may not be repeatable at
other locations.

Stage III: Screening for Symbiotic Effectiveness under Physical, Chemical, and Biological Stress of Site Soils

The fully effective N_2-fixation effectiveness expressed under
stage II conditions will not necessarily be upheld under real field
conditions. Thus, before selecting a final cluster of three strains
of <u>Rhizobium</u> for field evaluation, it is advisable to subject a
larger group (ten) of potentially effective strains to some of the
physical, chemical, and biological stresses of soils from the sites
for which the inoculant is being developed. This stage is particularly useful if the specific selection objective(s) includes adaptation to a particular stress, such as soil acidity in this case.
Stage III still has a value in selection programs, even for "non-
stress" soils. In stage II sand-jar evaluation, the test strains
did not have to compete against native soil rhizobia.

This third stage involves a pot experiment in which strains are
tested with the host plant, and the production is related to that of
inoculated control plants and N_2-fertilized plants. Soil is collected from the plow layer and mixed to uniformity to produce a homo-
geneous experimental material. Unsterilized soil is used. Soil may
be amended at fertilizer rates equivalent to field practice, but only
the N-control plants receive N (equivalent to 100 kg N/ha). Procedures for calculating the fertilizer additions are detailed elsewhere. Not all soils behave satisfactorily in pot experiments, and
other amendments may be necessary, particularly with heavier soils.
The following should be considered: (a) sieving to remove large

soil aggregates and stones; (b) addition of high C:N ratio residues such as bagasse at 1 to 2% (dry weight basis) to counter-balance excessive mineralization of N resulting from soil handling; and (c) addition of volcanic cinder, vermiculite, or other materials to improve soil aeration and drainage.

Sowing procedure and inoculation are the same for sand jars in stage II. About 6 to 8 seedlings are planted and thinned to 2 to 4 plants per pot, depending on the species. Thinning is by snipping off the plant tops, rather than by pulling entire plants from the soil. Pot size is optional, but 20 to 25 cm in diameter is usual. Six replications of each treatment are required.

Precautions against cross-contamination in this stage are essential. Watering, which in greenhouses in the tropics is needed daily, is the primary source of contamination. It can be minimized by filling pots so that soil level is 3 cm below the pot rim, watering gently to avoid splashing, using grid or mesh benches instead of solid benches so that pots can drip through onto the floor, raising pots on supports (such as petri dish lids) so that there can be no water flow on the bench surface from the emergent roots from one pot to those of another, and assigning watering to a single, informed individual. Other precautions include avoidance of overheating of the roots and nodules in pots and minimizing nontreatment effects. Pots should be set up in a randomized complete block design but not rerandomized thereafter because of the overriding problem of contamination through handling.

As with sand jars, plant dry-matter production is the most meaningful parameter to be determined and is the basis for ranking strains. The top three strains are promoted to stage IV.

Stage IV: Single-Location Evaluation of Strains of Rhizobium and Inoculation Methodology under Field Conditions

Strains emerging from stage III are evaluated for nodulation and N_2 fixation under field conditions. Although the preferred measure of the response by a legume to inoculation with the test strains of Rhizobium is grain yield (dry matter production in the case of forages), there are many factors which, under field conditions, can prevent differences in N_2 fixed by the strains being translated into differences in yield. Therefore, field trials should include a mid-season harvest to determine dry-matter production. Plot size should be sufficient to house two fully bordered harvest areas. The standard plot layout used in the International Network of Legume Inoculation Trials (Halliday, 1981a; NifTAL, 1980) is suitable (Fig. 4).

When the specific selection objective includes overcoming soil stress, the field trial at stage IV can amalgamate the strain selec-

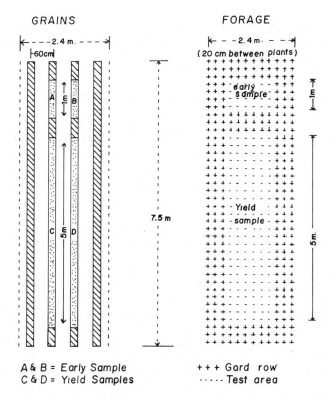

GRAINS   FORAGE

A & B = Early Sample   + + + Gard row
C & D = Yield Samples   ····· Test area

Fig. 4.   Plot design for field evaluation of a strain of Rhizobium
          (stage IV).

tion approach and other strategies for overcoming the stress.   In
this case, several inoculation methods were appraised for their ability to overcome the effect of acid soil stress on nodulation.   Simple
seed inoculation with an aqueous suspension of peat-based inoculant
containing the test strains was one treatment.   Others involved pelleting the inoculated seeds with finely-milled lime or rock phosphate.   These treatments were compared to control plots of uninoculated plots and plots fertilized with N.   The comparison between
these treatments is most valid, in a scientific sense, when there
are no other factors limiting plant growth.   But the comparison is
most realistic when the level of agronomic inputs is economically
feasible and similar to that used by farmers in the region where the
legume is grown.   In the procedure adopted, the scientific validity
was considered of lesser importance than the need to be realistic,
and a minimal blanket fertilization of elements other than N was applied.   Three replications of the treatments were established in a
randomized complete block design.   Experience has indicated that
four replications are desirable.

     Precautions against cross contamination are of paramount im-
portance.   Common pathways of contamination are careless handling of

inoculated seed at planting time, use of field implements without
sterilizing them between plots, tramping from plot to plot (by la-
borers, animals, visitors, etc.), and run-off and other drainage
problems caused by poor site selection.

The best Rhizobium-inoculant method combination is then selected and subjected to further testing in stage V. It could be justified to produce and use legume inoculants based on stage IV evidence,
but there remains the risk that the selected strain will be a successful inoculant only in the specific soil and climatic conditions
under which it was selected. A further stage (stage V) is essential
to determine the range of suitability of inoculants developed for a
single location in stage IV.

Stage V: Multi-Location Testing of the Response
to Inoculation with Selected Rhizobium Strains

A standardization design developed for the International Network of Legume Inoculation Trials (INLIT) is available for those
contemplating multi-location trials on the response of legumes to
inoculation with selected strains of Rhizobium (NifTAL, 1980). One
of the major constraints to fuller utilization of legume inoculation
in the tropics is that there has not been convincing demonstration
on a wide scale that yield increases will result with local legume
varieties under local soil and climatic conditions (Halliday, 1981a).
Stage V trials can assist in deriving the data necessary for pre-
dicting more reliably whether a legume will respond to inoculation
or not.

The trial has three basic treatments: plants inoculated with
Rhizobium, plants not inoculated, and plants not inoculated but fertilized with N.

The comparison between these treatments is most valid, in the
scientific sense, where there are no other factors limiting plant
growth. The comparison between these treatments is most realistic
when the level of agronomic inputs is economically feasible and
similar to that used by farmers in the regions where the legume is
grown. Therefore, the comparison is made at two fertility levels
which, for convenience, shall be referred to as "farm fertility" and
"maximal fertility."

Fertilizer levels are determined on the basis of information
available locally.

With three treatments at two fertility levels replicated four
times, a 24-plot, randomized, complete block design results. The
treatments in the first replication can be deliberately arranged to
serve as a demonstration in which the treatments that are most frequently compared are located side-by-side to facilitate visual ob-

servation of treatment differences. Plot arrangement is the same
as for stage IV (Fig. 4). Row spacing, planting distance, and seed
depend on the legume in question.

The plus N control plots receive 100 kg N/ha but in two doses.
At planting, 25 kg N/ha is applied, and 75 kg N/ha is added 4 to 5
weeks later in the case of grain legumes. With forages, the 25 kg/
ha is applied at planting, and 25 kg N/ha is applied after each cut
(at approximately 3-month intervals).

It is best to sow the "uninoculated" and "N-fertilized" plots
first. Only after the seeds in these plots have been covered are
the inoculated seeds prepared for sowing in the remaining plots.
This minimizes the risk of contamination of the plots that are not
to receive rhizobia.

Stage V trials can be used to characterize selected strains for
competitive ability and persistence if the inoculant strain is
"marked" serologically or with antibiotic resistance. Such strains
of Rhizobium can be detected in the nodule population and their ability to compete against strains native to the site determined. These
strains can also be detected, if present, in the soil in following
seasons, or in the nodule populations of subsequent legume crops
sown uninoculated.

The International Network of Legume Inoculation Trials coor-
dinated by the University of Hawaii NifTAL Project is available for
15 agriculturally important tropical legumes. Inoculants developed
for INLIT each contain three serologically distinct, effective
strains of Rhizobium from diverse geographic and host-germplasm
backgrounds (Halliday, 1981a; NifTAL, 1980). Each INLIT is therefore potentially an ecological study of the relative performance of
the three exotic strains between themselves and in competition with
indigenous soil strains and is also a long-term persistence trial.

ADDITIONAL SCREENING STAGES

For some specific selection objectives, the development of
rapid screening procedures may reduce the time taken to develop a
reliable inoculant strain, or may greatly increase the likelihood
of successful inoculant strains emerging from the step-wise screening previously described. In the case of selection of rhizobia for
acid, infertile soils, a laboratory prescreening that preceded the
stage I test greatly increased the range and numbers of strains that
could be addressed. It eliminated effective strains predestined to
fail in the field but which would have passed through stages I, II,
and possibly III, consuming time and resources. The prescreening
test was based on the reasonable assumption that for a strain of
Rhizobium to be a successful inoculant for legumes grown in acid

soils, the ability to multiply well at low pH is an essential trait.
Synthetic media were developed that tested the ability to multiply
at low pH, and only those strains passing the test were introduced
into the step-wise screening program (Date and Halliday, 1979).

Investigators may find it useful to develop such rapid prescreening steps appropriate to their own specific selection ob-
jective(s).

CONCLUDING REMARKS

As with any screening program, there is always the risk that
discarded materials that could not be accommodated in the later
stages would have performed well in the field. In the procedure
described, the stage-to-stage transition that is most problematic
is that from stage II to stage III. Rankings of strains in sand
jars do not necessarily hold up when subjected to the stresses of
site soils. Although 10 fully effective strains are passed across
from II to III, examples have occurred in which as few as three of
the strains could nodulate at stage III and only one of these was
effective (Halliday, 1981b).

When dealing with uncommon legume species, an investigator
should be concerned about whether the routine media used in stage I
and stage II are, in fact, nonlimiting to growth of the legume plant
so that Rhizobium characters can be expressed. As an example of
this, it was found that Stylosanthes capitata a legume with high
tolerance to soil acidity factors and native only to regions of
South America with acid soils, could not be nodulated by any one of
more than 100 Stylosanthes isolates (including many specifically
from S. capitata) tested in stage I. Nor would S. capitata grow in
stage II. Only when the growth medium was acidified to a pH value
of lower than 5.0 and the Ca and P levels lowered ten-fold would the
plant nodulate and grow.

Even though the screening procedure is lengthy, attempts to
shorten the sequence are ill-advised. Recommendation of strains
of Rhizobium for legume inoculation without first performing field
trials similar to those described in stage IV and stage V would be
foolhardy in the face of accumulating data that indicate that site
variation in performance of selected strains is common (Halliday,
1980).

REFERENCES

R. A. Date and J. Halliday, in: Handbook for the Collection, Preser-
     vation and Characterization of Tropical Forage Germplasm Re-
     sources, CIAT, Colombia, pp. 21-26 (1979a).

R. A. Date and J. Halliday, Nature (London), 277:62-64 (1979b).

R. A. Date and J. Halliday, in: Advances in Legume Science, Kew, England, pp. 575-601 (1980).

J. Halliday, in: Pasture Production in Acid Soils of the Tropics, CIAT, Colombia, pp. 123-127 (1979).

J. Halliday, Workshop on Agrotechnology Transfer, Benchmark Soils Project, University of Hawaii, Honolulu, Hawaii (1981a).

J. Halliday, in: Global Impacts of Applied Microbiology (S. O. Emejuaiwex, O. Ogunbi, and S. O. Sanni, eds.), Academic Press, New York, pp. 73-84 (1981b).

J. Halliday, in: Workshop on Innovative Biological Technologies for Lesser Developed Countries, Office of Technology Assessment, Washington, D.C., pp. 243-273 (1981a).

NifTAL, International Network of Legume Inoculation Trials, Experiment A, University of Hawaii NifTAL Project, Honolulu, Hawaii (1980).

NifTAL, International Network of Legume Inoculation Trials, Experiment B, University of Hawaii NifTAL Project, Honolulu, Hawaii (1982).

P. Somasegaran, H. Hoben, and J. Halliday, Practical Exercises in Legume/Rhizobium Technology, University of Hawaii NifTAL Project, Honolulu, Hawaii (1979).

CURRENT USE OF LEGUME INOCULANT TECHNOLOGY

P. M. Williams

Instituto Venezolano de Investigaciones Cientificas
Centro de Microbiologia y Biologia Cellular
Apartado 1827
Caracas 1010-A, Venezuela

INTRODUCTION

A constantly growing world population has resulted in an ever increasing demand for food. There were once great hopes that new agricultural techniques, primarily the development of high yielding grains, would solve the problem. Unfortunately the grains require large amounts of nitrogen fertilizer, the price of which has soared over the years because of the cost and scarcity of the fossil fuels required for its commercial synthesis.

Fortunately, there is a great potential to increase soil and plant N supplies without depleting the world's energy resources through biological N_2 fixation. The ability to fix N_2 resides with some free-living microorganisms as well as with others in symbiotic associations. Perhaps the best known relationship is between the bacterium, Rhizobium, and the leguminous plant. Although the amount of N_2 fixed by legumes varies with crop and environmental conditions, the average fixation is thought to be between 50 and 100 kg N ha^{-1} yr^{-1}. On a global scale, the N_2 fixed by plant processes has been estimated to amount to 90 million tons yr^{-1} (35 million by crop legumes, 45 million in permanent meadows and grasslands, and 9 million tons by nonleguminous plants) as compared to 50 million tons yr^{-1} fixed (manufactured) industrially.

In order to maximize biological N_2 fixation in agriculture, it is necessary to assure the participation of rhizobia with high N_2-fixing capabilities in the symbiotic association. This can be accomplished by rhizobial inoculation of leguminous seeds. The prime objective of legume seed inoculation is the nodulation of the host plant, and it consists of the introduction of a strain of <u>Rhizobium</u> into the soil at sowing in a form that enables it to (a) remain viable until the host seedling can be infected, (b) compete with any indigenous or naturalized rhizobia for infection sites and so form sufficient root nodules to permit maximum N_2 fixation in the host legume, (c) adequately nodulate its host(s) rapidly over a wide range of environmental conditions, and (d) persist in soil in numbers sufficient to maintain nodulation of perennial legumes or to achieve prompt nodulation of regenerating annual species.

The early history of legume seed inoculation involved soil transfer methods; i.e., the treatment of a sown area with large amounts of soil, up to 3 tons ha^{-1}, taken from around the roots of a good crop of the legume in question, or smaller quantities of such soil could be applied to seed in the form of a dust or paste (Fred et al., 1932). The practice of inoculating seed with pure cultures of rhizobia started in 1896 after the discovery of the ability of the inoculated legume to fix N_2 and the isolation of the rhizobia from the root nodule (Voelcker, 1896). The pure cultures were initially suspended in water and the suspension used to impregnate the soil or for seed inoculation. These early inoculants were soon replaced by sterilized soil impregnated with rhizobia, subsequently by peat coated with agar, and finally in the 1920's by peat alone impregnated with rhizobia (Fred et al., 1932).

Peat-base inoculants presently constitute the vast majority of cultures marketed today, and their development is primarily due to their convenience in holding and distributing the rhizobia. Such inoculants offer added advantages over agar, broth, or lyophilized cultures, these including increased protection against acid fertilizers (Vincent, 1958), better survival on the seed (Date, 1968), and improved survival beneath a lime pellet (Shipton and Parker, 1967), and generally solid-base inoculants result in better nodulation in the field than when other forms of inoculant are used (Brockwell and Whalley, 1962; Burton and Curley, 1965; Date, 1968; Shipton and Parker, 1967).

As we become more dependent on biological N_2 fixation, it is imperative that inoculants of proven quality be available. Inoculant technology is implemented on a commercial scale mainly in the developed countries, the United States and Australia having substantial industries for their production, distribution, and marketing. There is also commercial-scale production in Brazil, Uruguay, Argentina, India, and Egypt. Certain research centers such as CIAT (Centro Internacional de Agricultura Tropical) in Colombia, the University of Hawaii NifTAL Project and IVIC (Instituto Venezolano de Investigaciones Cientificas) in Venezuela produce inoculants in pilot-scale plants as a service mainly to researchers and occasionally to legume growers.

Demands for inoculant technology are increasing, this being mainly a result of the increased use of soybeans. There are dangers in trying to satisfy this demand by importation of inoculants developed in the United States or elsewhere through subsidiaries of the parent company or suppliers of legume seed. However, Rhizobium strains and inoculation methods for particular environmental conditions and farming methods are rarely transferable from one country to the next. Moreover, the storage conditions during shipment of legume inoculants may greatly affect rhizobial viability, thus invalidating the checks for quality control performed by the manufacturers prior to marketing.

Two basic techniques for the commercial production of peat cultures of Rhizobium exist. The Australian or United States system depends upon the impregnation of the peat with high-count broth cultures and is not dependent on rhizobial growth in the carrier. In the European system, the carrier is enriched with nutrient broth and autoclaved before being seeded with a small quantity of rhizobial culture; sufficient rhizobial multiplication occurs in the carrier to meet a minimum standard.

Various aspects of the commercial preparation of legume seed inoculants are discussed in this chapter, along with procedures for quality control and techniques for inoculant application.

Rhizobium STRAIN SELECTION FOR INOCULANTS

The selection of suitable strains of Rhizobium is basic to the process of inoculant production and commonly demands specific cultures for species, groups of species, or even varieties in the one inoculation group. Only the criteria for the selection of strains will be presented here. Vincent (1956) stated the basic criteria as (a) selected Rhizobium strains must promptly form effective nodules with appropriate host(s) and (b) selected strains must have the ability to do this under diverse field conditions.

Effective nodulation is still the main criterion used in strain selection, but the list of criteria has expanded with time to include the following ecological and agronomic characteristics, (c) the ability to form nodules in competition with indigenous or naturalized rhizobial populations, (d) the ability to survive in the absence of the host, (e) the ability for rapid and effective nodulation over the widest possible root temperature range, (f) the ability to grow in broth and peat culture, (g) the ability to survive in the inoculant carrier and on the seed, (h) the ability to resist variations in soil pH and to tolerate macro- and micro-nutrients and pesticides in close proximity, and (i) genetic stability in culture.

The procedure for strain selection normally involves a preliminary screening of several strains, using plants grown in bottle-
jar assemblies (Leonard, 1943), fully (Purchase and Vincent, 1949),
or partially enclosed tubes (Gibson, 1963) or growth pouches (Weaver
and Frederick, 1972) under greenhouse, light room, or growth-chamber
conditions. A small number of selected strains are then subjected
to more rigorous and more costly field trials. Strains presently
in use would normally be included in such a testing program.

The final selection of a <u>Rhizobium</u> strain for use in commercial
cultures is very often a compromise between a strain with excellent
performance with particular hosts and/or environments and a "wide-
spectrum" strain of good performance with the general range of hosts
under a wider range of environmental conditions. Alternatively,
multiple-strain inoculants containing the best strain for each host
species may be produced. In Australia, "wide-spectrum" strains are
used when these are available, but there is increasing use of specialized inoculants with specific strains for individual hosts.
Despite findings that suggest that multi-strain inoculants should
be avoided because of possible antagonistic and competitive effects
in culture and the likelihood of competition in nodule formation
from the less effective strains (Marshall, 1956; van Schreren et
al., 1954), this is the approach used successfully by the inoculant
industry of the United States.

Strains for testing can be obtained from other laboratories
working with the same species, or from nodules formed on the legume
by native strains after sowing uninoculated seed in the region where
the new species is expected to be used. None of these sources is
invariably better than the other in screening programs.

In some countries, strain selection and testing are largely the
responsibility of the producers, with some assistance given by
universities and other research institutions. This situation occurs in the United States, whereas in Australia, for example, the
responsibility lies with government and university research organizations to select, evaluate, and maintain strains that are then
made available to industry as required.

Maintenance of Strains

The bacteria in most major culture collections are normally
maintained as lyophilized cultures held at 5°C. This system maxim-
izes the storage period (15 to 20 years) and minimizes genetic
variation and contamination. When freeze-drying equipment is not
available, cultures may be dried on porcelain beads over a dessicant
in screw-cap bottles (storage period 3 to 4 years) (Norris, 1963).
Whichever system is employed, authenticity checks should be carried
out regularly. Screw-cap test-tube slope cultures are normally used
to hold cultures in a convenient form for short-term storage (1 year).

TYPES OF INOCULANTS

Agar Cultures

The first commercial inoculants, "Nitragin," were produced on gelatin and subsequently on agar nutrient media. Such inoculants were used either by direct application to the seed, using bacterial suspensions in skim milk, or by mixing the bacteria with soil or finely cut leguminous hay, which was subsequently spread over the field. Agar cultures were replaced by solid-based inoculants in the 1920's in the United States and in 1952 in Australia mainly because of the high mortality rate during drying following seed application. Mortality during the drying phase can be significantly reduced by the addition of 9% maltose to the suspending fluid (Vincent et al., 1962).

Liquid Cultures

Liquid cultures of Rhizobium are normally used for inoculation purposes in the laboratory and greenhouse and in the preparation of commercial solid-base inoculants. Broth cultures are not normally available as commercial inoculants except when manufacturers produce specific cultures on a one-off basis for research purposes or for specialized crop growers.

Lyophilized Cultures Carried in Talc

The objective of the lyophilization process is to remove the intracellular water from the rhizobial cell; this reduces metabolic activity to a minimum, thus increasing longevity. This is achieved by freezing the intracellular solutions and removing the moisture through the cell membranes under vacuum, without allowing the cellular contents to thaw. Lyophilization of broth cultures requires sophisticated equipment, technical expertise, and the use of buffers to enhance viability during drying of the cells and during reconstitution; the process is therefore expensive. Talc is often used as a carrier-filler to aid inoculation on a large scale. Such inoculants are not available commercially because of the high production costs and poor survival of the cells on the seed (Vincent, 1965).

Oil-Dried Cultures

High-count broth cultures are suspended in liquid oil of plant or mineral origin, and by bubbling air through the suspension, the water content of the cell may be reduced without affecting its viability. Moisture contents of from 1 to 10% are achievable, and at the lower level, cultures are claimed to be stable up to 73°C. After the drying process, the cells are collected by filtration or centrifugation and mixed with a carrier, such as oil or finely ground

vermiculite. The use of oil is designed for its direct application
to seed in a row-planter hopper. The use of vermiculite results in
a powder with an oily texture which facilitates application of the
powder to the seed without the need for an adhesive. Results obtained with this type of inoculant have been inconsistent (Hiltbold
and Thurlow, 1978; Schall et al., 1975).

Concentrated Frozen-Broth Cultures

High-count broth cultures are concentrated into a paste (approximately 1×10^{12} ml^{-1}), packaged, and then "snap" frozen. The
frozen cultures are distributed in plastic tubes, bottles, or can-
isters and are maintained in the frozen state by dry ice. They may
be stored at below 0°C for at least 9 months. Once thawed, the cultures should be used within 24 h. There has been a rapid increase
in the commercial use of this form of inoculant in the United States,
especially in the southern states, where environmental, soil, and
disease problems have affected the usefulness of conventional inoculants. Frozen-concentrated Rhizobium cultures are used only with
the spray-in inoculation technique (i.e., the culture is diluted in
water and applied to the soil), and such cultures are presently only
produced for soybeans and peanuts. Published results suggest that
under harsh conditions, good results can be obtained (Scudder, 1974).

Solid-Base Cultures

Peat and soils high in organic matter are the most common carrier materials for rhizobia in legume inoculants because of their
protective action against the adverse effects of high temperature
and low moisture content during storage and when the rhizobia are
added to soils. Efforts to obtain carriers superior to peat have
been directed to a neutralized soil-peat base enriched with nutrients. Among these are peat containing carbon black (Gunning and
Jordan, 1954; Newbould, 1951), carriers consisting of peat or soil
supplemented with materials such as alfalfa meal, ground straw,
yeast, and sugar (van Schreven, 1958, 1963, 1970; van Schreven et
al., 1954; Wrobel and Ziemiecka, 1960), soil plus coir dust or soybean meal (Iswaran, 1972; John, 1966), soil plus wood charcoal
(Gunning and Jordan, 1954; Iswaran, 1972; Newbould, 1951), and peat
amended with nutrients and an adhesive to improve its ability to
adhere to seeds (Hastings et al., 1966). Other carriers have been
used, with varying success, where peat deposits are unavailable or
are unsuitable. The alternative carriers include, for example,
vermiculite, decomposed sawdust, perlite, and rice husk compost
(Bonnier, 1960), milled, decomposed maize cobs supplemented with nutrients (in Zimbabwe) (Corby, 1976), and a coal-bentonite-alfalfa
meal carrier (in South Africa) (Deschodt and Strijdom, 1976; Strijdom
and Deschodt, 1975). Finely ground bagasse has been shown to be
suitable for soybean rhizobia (Leiderman, 1971), especially after
washing to remove excess sugar (P. H. Graham, personal communica-

tion). Autoclaved or gamma-irradiated filter mud has been shown
to support high numbers of clover, cowpea, and soybean rhizobia
during storage and when inoculated on seed (Philpotts, 1976; Stein
et al., 1980; Williams and Sicardi, 1980), and is currently used in
the pilot-scale preparation of inoculants in Venezuela.

The production of peat-base inoculants is detailed in the following section.

Granular Inoculants

These inoculants were developed as a direct implant form for
sowing into the row beside the seed and are of particular value under the following circumstances. (a) Conditions where seed inoculation is hazardous because of fragility of the seed coat. (b)
Certain epigeal species, particularly soybean and subterranean
clover, frequently lift the seed coat out of the soil during emergence of the cotyledons. In such cases, rhizobia would not be de-
posited in the soil if seed inoculation techniques were used. (c)
Circumstances when seeds are treated with insecticides or fungicides that endanger rhizobial survival. (d) With small seeded
leguminous plants, seed inoculation may not provide sufficient
rhizobia to effectively nodulate the plant. (e) Circumstances when
high levels of inoculant need to be applied to allow the inoculum
strain to compete with naturally occurring rhizobia.

Granular inoculants are prepared in the United States using
peat, clay, and occasionally vermiculite. Peat granules are naturally formed during the flash drying of peat. They are screened to
pass a 20-50 mesh and mixed with high-count broth cultures to raise
the moisture content from 7-8 to 32-34%. Such granules flow freely
through field-implantation equipment. Kaolin and sodium bentonite
are the principal clays used and the preparation of the inoculant
is basically the same as for peat granules. Vermiculite granules
(30-60 mesh) are mixed with oil-dried rhizobia and have a slightly
oily texture, but they flow freely through implant machinery. In
New Zealand, granular inoculants are made by the application of peat
inoculants in either a nutrient solution or inert carrier to a
marble chip.

Vacuum-Impregnated Inoculants

The process of seed impregnation with inoculants is commercially
known as "noculation," and this process may be considered as a form
of inoculation in which the seed acts as a carrier and the seed coat
offers protection to the rhizobia. Concentrated, frozen broth cultures (1×10^{12} ml^{-1}) are thawed, added to an appropriate weight of
seed, and mixed until the seeds are evenly moistened. The seeds are
transferred to a vacuum chamber for impregnation. On removal of the
vacuum, rhizobia are thought to be drawn into the seed through the

micropyle and into the pores, scratches, and cracks in the seed
coat. Little imbibition of the seed occurs during the impregnation process. Although the process offers advantages to the manufacturer and the farmer, rhizobial survival has been reported to be
poor (Date and Decker, 1962; Schall et al., 1975 .

COMMERCIAL PREPARATION OF PEAT-BASE INOCULANTS

Broth-Culture Preparation

Media. Most reseachers today produce rhizobial cultures in a
yeast-mannitol broth with the following composition: 0.5 g of
K_2HPO_4, 0.2 g of $MgSO_4 \cdot 7H_2O$, 0.1 g of NaCl, 10 g of mannitol, and 1.0 g of yeast extract in 1.0 liter of distilled water
(Burton, 1967; Fred et al., 1932; Roughley, 1970; Vincent, 1970).

Although rhizobia utilize mono- and disaccharides, slow growing strains such as R. japonicum and R. lupini prefer pentoses
(arabinose) to sucrose or mannitol (Graham and Parker, 1964). Despite these preferences, manufacturers of commercial inoculants
usually use sucrose to minimize production costs.

The production of cultures with a high percent viability is
desirable for inoculant preparation. Increased numbers of viable
cells can be achieved by optimizing the amount of yeast extract in
the medium for individual strains, care being taken not to inhibit
growth (Date, 1972; Sherwood, 1972) or cause cell distortion
(Jordan and Coulter, 1965). Yeast extract (1%, Difco, Oxoid,
Vegemite) in liquid media favors rapid growth and high percentages
of viable cells in cultures of R. japonicum CB 1809, R. lupini WU
425, R. meliloti SU 47, R. trifolii TA 1, and cowpea strain SB 756.
Concentrations of yeast extract greater than 0.35% depress viability
and produce distorted cells in all strains except SU 47, with TA 1
being especially sensitive (Skinner et al., 1977).

Fermentation. On a small scale, rhizobial cultures may be con-
veniently produced in cotton wool-plugged Erlenmeyer flasks that are
agitated at 28°C on rotary or reciprocating shakers. Larger scale
production requires fermentation equipment, which may range from
elaborate industrial fermentors of 1000 to 9000 liter capacity
(Burton, 1967; Date, 1965; Hendrickson, 1942), to glass or metal
vessels of 10 to 100 liter capacity (Skinner and Roughley, 1969;
van Schreven, 1958). A 20-liter glass bottle (containing 15 liter
of broth) and a 75 liter stainless steel vessel (containing 50 liter
of broth) used for the pilot-scale production of inoculants in the
author's laboratory is shown in Fig. 1.

Each fermentor is fitted with an air-inlet and an air-outlet
tube and with inoculation and sampling ports. The air-inlet tube

Fig. 1.   A.   20-liter glass fermentor.   B.   75-liter stainless-steel
          fermentor.   (a) Inlet tube fitted with filter for steril-
          ization of air.   (b) Outlet fitted with filter to prevent
          back contamination.   (c) Inoculation port.   (d) Sampling
          port.

extends to the base of the vessel to provide adequate stirring of
the broth.   Both air tubes are fitted with filters packed with cotton or glass wool, and the whole unit is placed in an autoclave for
sterilization for 1 h, during which time the outlet tube is left
open to allow for pressure changes.   Larger fermentors are fitted
with coils or dimple jackets for steam sterilization and cooling
purposes, and the use of fine pore spargers and/or impellers is normally required to ensure adequate solution of oxygen.   After cooling, the fermentors are attached to an air supply, and it is often
advantageous to withdraw a broth sample after 48 h of aeration to
check the sterility of the system prior to inoculation.   Because
vigorous aeration of broth cultures results in more rapid growth and
higher cell yields, the optimal air flow rate should be determined
for each strain.   Air flow rates in the literature range from 0.5
to 120 liter of air h^{-1} $liter^{-1}$ of medium but 5 liter h^{-1} $liter^{-1}$ is
normally adequate (Roughley, 1970; van Schreven, 1958).

     The fermentors are inoculated at a level (0.1 to 1%) that gen-
erally provides 10^6 to 10^7 rhizobia ml^{-1} of culture medium at the
beginning of the fermentation.   Incubation temperatures normally are
26-28°C, but higher temperatures are sometimes used in commercial
fermentation processes (Burton, 1967).   Fast growing strains (generation times of 2 to 4 h) reach maximum viable numbers in 2 to 3
days, whereas slow growing strains (generation times of 6 to 12 h)
require 5 to 8 days.   Following fermentation, the broth should not

be stored for more than 24 h, as reductions in viable numbers may
occur (van Schreven, 1958). If broth cultures need to be stored,
this should be at 4°C.

In Czechoslovakia, a semi-continuous system (the residue from
one batch serving as the inoculum for the next fermentation) rather
than batch fermentation is used in large-scale broth culture preparation and the use of penicillin (100 units ml^{-1}) to reduce growth
of contaminating organisms has been reported (Weaver and Fredrick,
1972).

Carrier Preparation

Suitable carriers for rhizobia in legume inoculants have the
following properties: (a) high water-holding capacity, (b) serve
as a nutritive medium for rhizobial growth, and (c) favor rhizobial
survival during inoculant storage and distribution and when inoculated onto seeds.

Peat and soils with a high content of organic matter are the
commonest carriers, and since their suitability may vary between deposits or within the same deposit (Roughley and Vincent, 1967; van
Schreven, 1970), their ability to support large rhizobial populations of particular strains requires testing prior to their use in
inoculant preparation. Most peat deposits are too acidic for use
as carriers without prior adjustment to pH 6.5-7.0, $CaCO_3$ being the
most satisfactory neutralizing agent. This is normally added to the
peat before milling to ensure efficient mixing, but in the case of
the Nitragin Company, of Milwaukee, Wis., United States, the neutral-
izer is added during the mixing of the peat and broth culture (see
Fig. 3).

It is now generally accepted that rhizobia grow and survive
better in sterile than in nonsterile peat, the extent of the difference being dependent on the suitability of the peat and the temperature and duration of storage. Studies have shown that for the preparation of consistently high quality cultures of slow growing rhizobia
(e.g., those for cowpea, lupin, soybean, and Lotononis), the use of
sterile peat is essential; its advantages, however, may be marginal
in the case of the faster growing species (Roughley and Vincent,
1967). The methods for sterilizing peat include flash-drying (e.g.,
Nitragin Co.), autoclaving (e.g., the Netherlands, Czechoslovakia,
and New Zealand) and gamma-radiation (e.g., Australia), the choice
being dependent on the type of packaging, the amount of cultures to
be prepared, and the available sterilizing facilities.

Flash-Drying. Freshly harvested peat is drained, screened for
large solids, and shredded on a conveyer system leading to the dehydrator. Wet peat enters the dehydrator and passes between an inner, internally heated cylinder and an outer, revolving cylinder

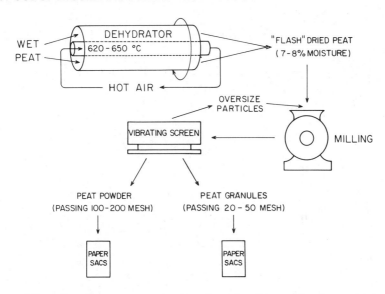

Fig. 2.   Peat preparation by the Nitragin Co.

(Fig. 2), the inner cylinder being heated by recirculated air with
an inlet temperature of 620 to 650°C.

The maximum time taken to flash dry the peat is 2 to 3 min, and
the water content is reduced from holding capacity to 7 to 8%. Although the process almost completely sterilizes the peat without
significantly denaturing its chemical components, there are ample
opportunities during the many subsequent handling points for contamination. The dried peat is ground in a hammer mill and passed
through vibrating sieves, which partition the material as required.
The particle sizes produced are fine powder (passing 200-300 mesh),
regular powder (the most common product, passing 100-200 mesh), and
granules (passing 20-50 mesh), these being formed naturally during
the tumble-drying action of the dehydrator. The very fine granules
remain intact during milling and sieving provided that appropriate
screens are used.

Autoclaving. Autoclaving has the advantage of being the only
method that allows absolutely pure cultures of rhizobia to be prepared and is used to sterilize peat for the "European" method of
inoculant production. Freshly harvested peat is drained in the field
and dried to a moisture content of 5% using forced-draught air. Drying temperatures vary but are below 100°C, higher temperatures causing toxic degradation in the peat and excessive increases in temperature when the peat is wetted with broth culture. Both of these
effects may restrict the subsequent growth and survival of rhizobia
(Roughley and Vincent, 1967). The pH of the dried peat is adjusted
by the addition of lime, and the peat is ground in a hammer mill be-

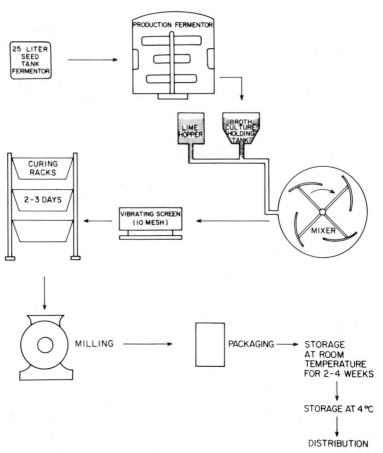

Fig. 3.   A schematic representation of the inoculant production
technique performed by the Nitragin Co. using flash-dried
peat as carrier.

fore passing to a vibrating screen. Peat powder, passing a 200-
mesh screen, is enriched with nutrient broth in a batch mixer prior
to autoclaving (Fig. 4). In the Netherlands and Czechoslovakia,
glass bottles, two-thirds full with moist peat, are plugged with
cotton wool and sterilized for 3 h at 121°C (van Schreven, 1958).
In New Zealand, pH-adjusted enriched peat is autoclaved and placed
in open containers in an aseptic growth chamber for subsequent seeding with a rhizobial culture (Fig. 4).

    Gamma Radiation.   In Australia, peat powder is packed in low-
density polythene bags (0.05-mm gauge) and sterilized by gamma rays
at a dose of 5 Mrads. Although gamma-irradiation does not completely sterilize the peat, the method is generally superior in pro-

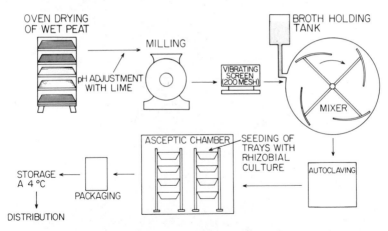

Fig. 4.   A schematic representation of inoculant production in New
          Zealand using the "European method." Autoclaved, pH-
          adjusted enriched peat is seeded with a culture of <u>Rhizo-
          bium</u> within an aseptic chamber.

moting growth of rhizobia to autoclaving peat for 4 h at 121°C
(Roughley and Vincent, 1967). A similar method is used for the
pilot-scale production of filter mud-based inoculants in Venezuela
(Stein et al., 1980).

Impregnation of Peat with Broth Culture

     Nonsterile Carrier. The following process is based on that of
the Nitragin Co., but in other countries where unsterilized peat is
used (e.g., in Uruguay), the procedure is essentially the same. At
the completion of fermentation, the broth cuture is transferred to
overhead holding tanks and is amended with calcium carbonate (315
liter of broth receiving 32 kg of $CaCO_3$) as it passes to a batch
agitator containing peat powder or peat granules (Fig. 3). A dia-
phragm pump is used to spray the broth, at low pressure, through
nonblocking, overlapping nozzles, and the broth is mixed with the
peat by rotating paddles, increasing the moisture content from 7-8
to 35-40% for powdered peat and to 32-34% for granules. The lower
moisture content of granules is essential in order to preserve their
discreteness. The peat-broth mixture is withdrawn from the mixer,
sieved through a 10-mesh vibrating screen, and matured in curing
racks for 2 to 3 days, during which time the heat generated through
wetting is dissipated. To prevent moisture loss from the peat surface, the curing racks are covered with a polythene film. At the
completion of curing, the peat is finely milled and packaged in polythene bags, which are sealed and punched with small ventilation holes
to facilitate gas exchange for subsequent culture growth and to expel
excess air during bulk packing. The inoculants are stored at room
temperature for 2 to 4 weeks, during which time maximum numbers de-

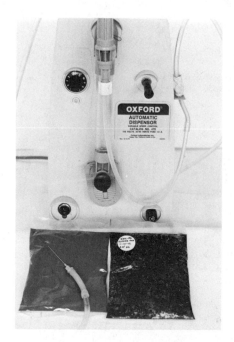

Fig. 5.   Impregnation of gamma-irradiated peat with broth culture.
Each packet is punctured with a hollow needle, and the required quantity of broth is delivered by an automatic dispenser connected directly to the fermentor. The puncture
is sealed with an adhesive label, and the contents are
mixed by hand.

velop and then they are stored at 4°C prior to distribution.

Granular inoculants are handled in a similar manner, except for
the sieving and milling stages, and are packaged in 32 kg polythene bags.

Sterile Carrier. The method for impregnating sterile peat without introducing contaminating organisms depends on how the inoculant
is packaged. When glass bottles are used, rubber tubing is used to
connect a hollow needle to the sampling port of the fermentor. The
flamed needle is inserted between the cotton wool stopper and the
neck of the bottle, and sufficient broth is siphoned over. The bottles are shaken after 24 h and incubated for 7 days at 30°C, after
which time the tops are covered with cellophane to reduce desiccation
and the inoculants are stored at 2°C. Counts of 4×10^9 g^{-1} for
clover cultures and 30×10^9 g^{-1} for medic cultures have been reported after the incubation period (van Schreven, 1958).

In New Zealand, the "European" method of inoculant production
is employed. Autoclaved, pH-adjusted enriched peat is held in open

trays within an aseptic chamber. The peat is seeded with a culture, and growth is maintained until a population of approximately 1×10^9 g^{-1} is achieved. The inoculant is packaged in polythene bags, which are sealed and stored at 4°C prior to distribution (Fig. 4).

Gamma-irradiated peat in polythene packets is inoculated by puncturing the film with a hollow needle and injecting broth with an automatic dispenser connected to the fermentor (Fig. 5). Sufficient broth is added to raise the moisture potential to pF range 2.69 to 3.42 (50 to 60% for Australian peat) (Roughley and Vincent, 1967). The puncture is sealed with an adhesive tape, the contents are mixed by hand, and the packets are incubated at 26°C for 2 weeks and then stored at 4°C.

Multistrain Inoculants

Such inoculants contain either a mixture of strains from two inoculation groups or a mixture of strains from the one group. The first type is produced to simplify distribution but obviously results in an inoculant with reduced numbers of rhizobia for each host group. The second type is intended to provide an inoculant effective for a wider range of host plants or to offset the danger of an undesirable mutation of one strain and for possible protection against bacteriophage attack.

Multistrain inoculants have the disadvantage that when two or more strains are grown together in mixtures of sterilized soil and peat (van Schreven et al., 1954), in broth culture, or sterilized peat (Marshall, 1956), one strain may become dominant in the mixture because differences in growth rate or antagonist effects.

Inoculants prepared using nonsterilized peat should be prepared by blending separate peat cultures immediately prior to packaging. Multistrain cultures in sterilized peat may be prepared by simultaneously injecting broths of each strain into the one container of peat.

Factors Affecting Rhizobial Numbers in Peat-Base Inoculants

Moisture Content. Both the initial and final moisture contents of peat cultures have marked effects on rhizobial numbers (Hedlin and Newton, 1948; Roughley and Vincent, 1967; van Schreven et al., 1954, Vincent, 1958), the optimal level differing for inoculants prepared with sterilized or nonsterilized peat (Roughley and Vincent, 1967). Because of the differing water-holding capacities of peats, the preferred method of expressing moisture is as moisture potential (pF) or bars (Fawcett and Collis-George, 1967) rather than as a percentage of the wet or dry weight of carrier. In nonsterilized Australian peat (Badenoch peat), pF values of 4.5 to 3.42 (40 to 50% moisture) proved optimal for a clover, an alfalfa, and a cowpea strain (Roughley and Vincent, 1967). The clover and alfalfa strains survived poorly at a pF value of 2.69 (60% moisture), and all strains died rapidly and the proportion of contaminating organisms increased at pF 4.88 (30% moisture). In sterilized peat, pF values of 4.15 to 2.69 were optimal, but the numbers of all three strains were restricted by a pF value of 4.88 after storage for only one week at 26°C.

Aeration. There have been conflicting reports on the requirements for gaseous exchange for growth and survival of rhizobia in peat cultures. Certain authors claim improved survival under conditions of free access of air (Hedlin and Newton, 1948; van Schreven et al., 1954), whereas others claim satisfactory survival in screw-capped jars or sealed cans (Gunning and Jordan, 1954; Newbould, 1951; Spencer and Newton, 1953). The growth and survival of clover, medic, and cowpea-type rhizobia were compared in sterilized peat contained in cotton wool-plugged tubes, sealed cans, and plastic film bags with various gas-exchange properties (Roughley, 1968). The sealed cans and polythene film of very low gas exchange were unsatisfactory, whereas rhizobial growth and survival were good in low and medium weight polythene film bags and comparable to that in plugged tubes. The practice of punching ventilation holes in plastic bags is considered unnecessary and harmful because it enhances moisture loss and allows the entry of contaminants.

Storage Temperature. The effect of storage temperature on the growth and survival of rhizobia is influenced by the purity of the culture and moisture loss during storage. When cultures are prepared in sterilized carriers, incubation at 26°C following impregnation promotes the initial rapid growth of rhizobia and has little effect on long-term survival if the moisture content is maintained (Roughley, 1968). Continuous storage at 4°C restricted the initial multiplication and maximum numbers developed after 26 weeks (Roughley, 1968). The survival of clover rhizobia in cultures of unsterilized peat stored at 25°C was poorer than those stored at 5°C (Vincent, 1958).

QUALITY CONTROL OF INOCULANTS AND PRE-INOCULATED SEED

The principal reason for quality control of inoculants and pre-inoculated seed is consumer proection, and although the minimum standards and evaluation methods vary from one country to the next, most quality control systems are based on the following principles: (a) to achieve an acceptable minimum number of viable rhizobia per seed, as a basis for a recommended rate of application; (b) to achieve an acceptable minimum population in the inoculant at purchase; and (c) to achieve an acceptable minimum population in the culture at the expiration date.

In some countries, for example, the United States, manufacturers generally have complete responsibility for inoculant quality (Burton, 1967), although some testing is carried out by state departments of agriculture of products purchased in retail outlets. In some countries, for example, Australia (Date, 1969), Holland (van Schreven, 1958), Uruguay (Sicardi de Mallorca and Labandera, 1979), U.S.S.R and Czechoslovakia (Hamatova and Vintikova, 1963), inoculant production is completely under the control of government agencies. The minimum standards and control practices of different countries are outlined below.

Australia

The Australian Inoculant Research and Control Service (AIRCS) (Thompson, 1976) is a group funded by state and federal governments as well as from industrial levies and is staffed by government employees. Although the AIRCS does not have legal powers to enforce the use of recommended strains, the submission of inoculants for testing, or acceptance of directives concerning expiration dates, batch numbers, or storage conditions, it functions on the basis of mutual acceptance of responsibilities by all concerned. AIRCS provides the following services: (a) supply of tested, recommended mother cultures to producers on request; (b) serological typing, purity, and checks of viable counts on inoculum batches; (c) quantitative and qualitative testing of peat cultures following their impregnation with broth cultures; and (d) quantitative and qualitative testing of peat cultures before and after their distribuiton to commercial outlets.

The Australian standard set for peat cultures is 1×10^9 viable rhizobia per gram of moist peat at manufacture and 1×10^8 per gram at expiration, with 0.1% contamination for cultures produced in sterilized peat. Because strains for Lotononis cannot achieve such levels in peat culture, standards of 3×10^8 and 3×10^7 per gram for manufacture and expiration dates are accepted. The expiration date, which is marked on the inoculant package alongside the batch number, is 6 months following withdrawal from cold storage. Inoculants may be kept for up to 6 months at 4°C prior to marketing.

Apart from the quality control testing of inoculants, AIRCS carries out research on aspects of inoculant production and use.

New Zealand

An official inoculant control system was initiated in New Zealand in 1979 and is known as the Inoculant and Coated Seed Testing Service (ICSTS) (Hale, 1979). ICSTS is operated by the Ministry of Agriculture and Fisheries and provides a similar service to AIRCS in Australia. The standard set for peat cultures is 1×10^8 rhizobia per gram up to the expiration date, which is 6 months, normally with an added allowance for cold storage. Inoculated seed must provide a minimum of 300 viable rhizobia per seed after storage at 20°C for 28 days and must give satisfactory establishment in field trials. The date of preparation of inoculated seed must be stated on the package, along with a recommendation that seeds should be sown within 6 weeks after inoculation.

United States

Quality control in the United States is largely left to the manufacturers and to the consumer apart from the states of Indiana, Wisconsin, Ohio, and Maryland, which have statutes which require that inoculants and pre-inoculated seed reach a certain standard. Although the standards vary from state to state, all are based on so-called "grow-out" tests. In Indiana, for example, the control system is based on the Indiana Legume Inoculant Law of 1937 (Schall et al., 1975), and the Indiana State Chemist is responsible for the testing of materials covered by the act. The testing procedure for laboratory-inoculated or pre-inoculated seed involves their growth in sterile sand or vermiculite for 4 weeks under controlled conditions, followed by determinations of nodulation and plant dry weight. If 90% or more of the plants have one or more nodules on or near the primary root and plant growth characteristics are comparable to inoculated control plants, the test is considered satisfactory. If 67 to 90% of the plants contain one or more nodules, the sample is considered fair. Samples giving less than 67% nodulated plants are considered unsatisfactory. The acceptable level of viable rhizobia per gram of inoculant can be extremely low as based on this test.

Canada

Agriculture Canada, the body responsible for quality control of inoculants in the country, specifies that the following information must be on the inoculant package: (a) name of legume(s) for which the inoculum is effective; (b) date beyond which the product should not be used; (c) the species of Rhizobium; (d) the number of viable cells per gram (minimum standard of 1×10^6 per gram); (e) net weight of inoculum; (f) batch number; (g) directions for use and application rate (to provide at least 1×10^3 viable rhizobia per seed); and (h) manufacturer's name and address. The number of viable contaminants in inoculants or on pre-inoculated seed must be at a level that does not affect viability or performance of the desired species.

Uruguay

The Laboratory of Soil Microbiology and Inoculant Control of the Ministry of Agriculture and Fishing was established in 1963 and controls the quality of all imported and national inoculants (Sicardi de Mallorca and Labandera, 1978). This laboratory provides the fol-

lowing services: (a) the supply of selected strains of Rhizobium
to manufacturers for the commercial production of inoculants; (b)
purity and viable count checks on inoculum batches (minimum standard
of 5×10^8 ml^{-1}), and (c) checks of viable counts on inoculant
batches before packaging and distribution (minimum standard of
1×10^7 g^{-1} at expiration).

The minimum standards for inoculants produced in other coun-
tries are: (a) Czechoslovakia, 3×10^8 g^{-1}; Holland, $4-25 \times 10^9$
g^{-1}; India, 1×10^8 g^{-1} at manufacture and 1×10^7 g^{-1} at expiration; and U.S.S.R., $5-10 \times 10^7$ g^{-1}.

Testing Methods for Inoculants and Pre-Inoculated Seed

Careful sampling techniques and handling of inoculants, pre-
inoculated seed, or broth cultures are essential for a successful
testing program. Labelled samples should be transported to the
laboratory in a cooled container and stored at 4°C as soon after
collection as possible.

Broth Cultures. Tests performed on broth cultures should be
carried out as soon as possible after collection because storage
may result in death of rhizobia (van Schreven, 1958). The tests
are designed to check the purity of the culture and to determine
the number of viable specific rhizobia. When the culture is intended for the impregnation of a nonsterile peat, high viable numbers will be essential, and a slight degree of contamination may
not be detrimental. For the seeding of sterile peat, a lower initial count may be permissible, but any contamination could offset
the advantage of prior sterilization. Tests to determine total
rhizobial numbers include turbidity and microscopic counts using a
Petroff-Hausser chamber. The total viable rhizobial numbers may be
determined by the plate-count method, and inclusion of congo red in
the plating medium may reveal contamination (Rhizobium strains absorb very little of the dye when grown in darkness, whereas some
contaminants absorb the dye strongly and produce red colonies). A
preliminary check of the purity of cultures is normally performed
using a gram-stained smear (Vincent, 1970) and by pH determination. Rhizobium is gram-negative and appears clear red under the
light microscope, whereas any gram-negative contaminants appear dark
violet. Plating cultures on peptone agar also serves as a rapid
check of purity. Rhizobium grows poorly on this medium and causes
little change in pH. Hence, growth with a marked pH change within
24 h is indicative of the presence of contaminating organisms.

Serological typing of broth cultures with specific antiserum
serves to determine strain identity. The methodology for the development of antisera and the serological reaction, either agglutination or precipitation, has been described elsewhere (Vincent,
1970). The inclusion of a plant-infectivity test, using an appro-

priate range of dilutions (10^7 to 10^{10}), although not quite as precise as a plate count for enumeration, provides a check on maintained symbiotic capability of the strain.

Peat-Base Inoculants. The determination of the number of viable rhizobia in peat-base or other solid-base inoculants is the
most important control check point in determining inoculant quality.
Although freshly prepared inoculants using nonsterile peat should
contain sufficient rhizobia to permit valid plate counts, a plant-
infectivity test is routinely carried out in any high quality testing program. Comparison of biomass or increases in plant N content
over uninoculated, N-free control plants provides useful information about the efficiency of the inoculant strain.

When the plate-count method is used, care is needed in distinguishing rhizobial from nonrhizobial colonies, and when the latter organisms are very numerous, they may overrun the plate and
lower the apparent rhizobial count. The addition of congo red to
the basal medium will aid rhizobial identification. The growth of
contaminating organisms may be suppressed by supplementing the basal
medium with oligomycin (30 µg/ml), rose bengal (7.5 µg/ml), and
pentachloronitrobenzene (50 µg/ml). In the latter medium, rhizobia
develop as white, gummy colonies when the plates are incubated in
darkness. When additives are necessary, plates of pure cultures of
the particular rhizobia being counted should be made on both the
basal and modified medium to determine whether the additives ad-
versely affect rhizobial growth and to aid rhizobial colony identification. Tests of strain authenticity may be carried out by culturing from colonies in the plate and the application of a serological test.

The number of viable rhizobia in inoculants prepared with
sterile peat may be readily determined by the plate-count method.
The status of contaminating organisms may also be evaluated, and if
contaminants are present in significant numbers, rejection of the
inoculant is warranted.

Pre-Inoculated Seed. The determination of the number of viable
rhizobia on inoculated seed is difficult by the plate-count method
since the numbers of rhizobia are lower and the numbers of contaminating organisms are higher than in inoculants. Although the use
of a differential medium may aid rhizobial identification and counting, a plant-infectivity test is the preferred method.

The Plant-Infectivity Test

Representative samples of broth culture, inoculant, or pre-
inoculated seed are diluted with saline solution (0.85%), and standard small inocula are added to aseptically grown seedlings growing
in appropriate culture vessels. After 4 to 5 weeks of growth, the

extent of nodulation is recorded. From the number of nodulated
plants for each dilution factor, the most probable number (MPN)
(Date and Vincent, 1962) of infective rhizobia in the sample may be
estimated by reference to probability tables (Fisher and Yates,
1963). Generally 10-fold dilutions are adequate for the estimation
of rhizobial numbers, and the dilution series should be extended to
a point where no rhizobia are likely to occur. The dilutions of
peat-base or other solid-base inoculants or broth cultures are prepared by the addition of 10 g or 10 ml, respectively, to a serile
dilution bottle containing 90 ml of saline solution. The addition
of glass beads to the first bottle will assure a uniform suspension
of the inoculant sample after shaking 100 times. A 10-ml aliquot is
transferred from one bottle to the next, and the contents are shaken
thoroughly after each transfer. A 1-ml aliquot from the 10th bottle
represents a 10^{-10} dilution.

Oil-dried cultures require the addition of surfactants, such
as Tween 85 or Span 85, to the saline solution in order to achieve
a representative suspension for subsequent dilution.

Pre-inoculated seeds are vigorously shaken with 1 liter of
sterile diluent, and the suspensions are serially diluted as previously described. Seed grinding has been suggested as a more effective way of recovering rhizobia which may be within the cracks in
the seed coat (Vincent, 1970).

EVALUATING THE NEED FOR INOCULATION

Legume root-nodule bacteria are widely distributed in the
world's soils because of the natural distribution of the Legum-
inoseae and the cultivation of leguminous crops. Despite this, specific rhizobia are absent or sparse in many soils, or the local naturally occurring population may be ineffective or sub-maximal in
its N_2-fixing capacity; under such circumstances, a beneficial response to seed inoculation would be anticipated. There are numerous reports demonstrating the need for inoculation, the evidence
being provided by poor nodulation and low yields or eventual crop
failure. Such diagnoses are expensive and time consuming, and more
efficient means are available.

Field Evaluation

The existence of naturally occurring rhizobia in the field may
be easily detected by inspection of the roots of the legume growing
at the site. The presence of nodules having pink or red interiors
is indicative of an effective association, whereas nodules with white,
green, or black interiors are ineffective. The formation of many
large nodules near the crown of the root normally indicates a high
population of competitive rhizobia, whereas small nodules scattered

on the lateral roots may be the result of late nodulation because
of low populations of rhizobia of low effectiveness (Vest et al.,
1973).

The physical appearance of the host plant is also a useful diagnostic. A healthy, green legume growing in a N-deficient soil is
normally indicative of adequate nodulation by effective rhizobia.
Chlorotic, stunted plants may result from inadequate nodulation, but
stress conditions or disease often produce similar symptoms; therefore, nodulation should always be examined. Unthrifty legumes with
well nodulated roots may indicate infection by ineffective rhizobia
or that some other environmental factor is limiting growth. In such
cases, an investigation of the weather pattern during crop development may prove useful. Periods of drought, for example, can reduce
N_2 fixation by legumes (Engin and Sprent, 1973; Sprent, 1971, 1972)
and, in some species, result in nodule senescence (Wilson, 1931,
1942). If the legumes are not thrifty and the roots are not nodulated, infective rhizobia may be absent or in insufficient numbers
in the soil or other factors, such as pH, may be limiting the crop,
the bacteria, or both.

The history of the field should also be taken into account when
assessing the need for inoculation. If a N-fertilized crop was
grown in the field in the previous season, residual N may be present
in sufficient quantities to inhibit nodulation (Gibson and Nutman,
1960; Munns, 1968; Oghoghorie and Pate, 1971) of the legume in question. If, however, unfertilized crops had previously been grown at
the site for several seasons, a vigorous crop would indicate effective nodulation.

When evaluating a natural stand, the appearance of associated
species, such as grasses, will often indicate the N-supplying power
of the soil. Areas of healthy green grass growing some distance
away from legumes suggest an adequate basal N supply. Under such
circumstances, the condition of legumes, nodulated or not, must be
attributed in part to this native N level. Unthrifty grasses may
reflect an inadequate N supply, or other factors, such as nematodes,
may be operative.

When few or no effective nodules are found on introduced or
native legumes, a response to inoculation would be expected. However, when some effective nodules are formed by the native rhizobial
population, the value of inoculation can only be determined by field
trials. In such cases, uninoculated plots will serve as a visual
check in the evaluation of the inoculum strain and effectiveness of
the native rhizobial population. Uninoculated plots supplied with
fertilizer N will serve to determine whether or not environmental
factors other than N are limiting crop yield.

Laboratory Evaluation

The presence, frequency, and effectiveness of naturally occurring rhizobia can be detemined in the laboratory by plant infection methods using bottle-jar assemblies (Leonard, 1977), fully
(Purchase et al., 1949) or partially enclosed tubes (Hamatova et
al., 1973) or growth pouches (Wilson, 1931).

METHODS OF INOCULATING LEGUMES

Seed Inoculation

The method used to inoculate seed must provide sufficient viable rhizobia to ensure adequate nodulation of the host plant. In
many instances, this can be as few as 100 rhizobia per seed, but
in case of severe environmental stress, numbers as high as 10,000
or even 500,000 per seed may be necessary (Brockwell, 1977; Brockwell et al., 1980; Date, 1970). Peat-base inoculants were initially mixed with dry or moistened seed immediately prior to sowing,
but because of weak adherence and poor rhizobial survival, nodulation occurred only under favorable conditions; the method is not
therefore recommended. Larger quantities of inoculant can be attached to seed by the use of an adhesive such as 10% sucrose (Vincent,
1958) or 40% gum arabic (Brockwell and Whalley, 1962). The latter
substance has been shown to be superior to 5% methyl ethyl cellulose
(Cellofas A) (Brockwell, 1962) and other substituted cellulose compounds (Date et al., 1965) in promoting rhizobial survival and has
the additional advantages of offering protection against the effect
of toxic seed coats (Vincent et al., 1962), promoting early nodulation, and enhancing growth of the plant (Subba Rao et al., 1971).
Certain sources of gum arabic contain preservatives which are lethal
to rhizobia.

Whether dry inoculum is shaken over moist seed or an inoculum
paste (slurry) is added to dry seed, complete mixing is essential.
It may be advantageous to add half the required inoculum to the
seed, mix, add the remaining half, and then mix again in order to
guarantee adequate coverage. On examination, each seed should have
black specks on it, indicating the presence of the carrier.

Seeds may be inoculated on a clean patch of concrete, in large
tubs, or in motor-driven rotating bowls, the chosen method depending
on the quantity of seed to be sown. Inoculated seed should be allowed to dry in a cool place (not in direct sunlight) and planted
on the same day. If a delay in planting of more than two days occurs, reinoculation is advisable.

Seed Pelleting

Pelleting of seed with finely ground (passing 300 mesh) coating
materials at near-neutral pH values is now a standard agricultural
practice used to protect rhizobia on the seed coat against the deleterious effects of sunlight, low soil pH, desiccation, acidic
fertilizers, and fungicides and insecticides often used to pretreat
seeds. Other advantages include claims that seed coating improves
germination (Lowther, 1977) and improved flight or ballistic characteristics of seeds of pasture legume species used for broadcast
and aerial oversowing.

Low pH has often been implicated in legume and inoculation
failures under temperate conditions, and therefore limestone has
been commonly used to form the pellet (Brockwell, 1977; Loneragan
et al., 1955; Norris, 1971a, 1971b, 1973a). However, since trop-
ical-legume rhizobia may not have pH or lime requirements like those
of temperate species, plus the demonstrated need for adequate phosphorus levels, the use of rock phosphate has increased (Norris,
1971a, 1971b). Other coating materials that have been used include
bentonite, bauxite, talc, gypsum, dolomite, calcium silicate, and
kieselguhr (Norris, 1972, 1973b). In areas where pH is very low or
where soluble Al and Mn present problems, the use of finely ground
limestone may be justified. However, since lime has an alkaline reaction that is lethal for rhizobia, the alkalinity varying from
source to source, rock phosphate may be considered as the material
of choice unless a definite advantage for limestone can be demonstrated. Bentonite is sometimes included by some manufacturers to
improve moisture absorption by, and retention in, the coating once
it is wet, and activated charcoal may be used to nullify the influence of toxic substances in the seed coats of such species as
Centrosema pubescens and Trifolium subterraneum. Some success has
been achieved in New Zealand, Australia, and the United States in
improving establishment and early growth by the inclusion of micronutrients such as Mo in the pellet.

Although pelleting is a simple on-farm technique (Brockwell,
1977; Norris and Date, 1976), custom-pelleted (by seedsmen at the
farmer's request) and pre-inoculated seed are now more popular, and
such pelleted material can be stored for 2 to 12 weeks, depending on
ambient conditions and the legume species.

Most pre-inoculation procedures are based on multiple coatings,
alternately of adhesive and finely ground pelleting materials as
used in simple pelleting. The peat inoculant is included as one or
more of these coating layers. After coating, the pelleted seeds are
rapidly dried to 6 to 8% moisture using vacuum refrigeration dryers
(Taylor and Lloyd, 1968) to prevent seed imbibition. This is a very
critical stage which is necessary to circumvent premature seed germination or seed putrefaction, and such drying largely determines the
quality of the product.

Direct Implant Inoculation

An alternative to pelleting and pre-inoculation in recent years has been the use of concentrated liquid or solid granular peat cultures (as described above) which can be sprayed or drilled directly into the soil with the seed during planting. These methods have been especially successful for introducing inoculant strains into situations where there are large populations of competing naturally occurring rhizobia in the soil (Bezdicek et al., 1978) or in cases of adverse conditions, such as hot dry soils (Scudder, 1974) and where insecticide or fungicide treatment of seeds precludes direct seed inoculation (Brockwell et al., 1980).

REFERENCES

D. F. Bezdicek, D. W. Evans, B. Abede, and R. E. Witters, Agron. J., 70:865-868 (1978).

C. Bonnier, Ann. Inst. Pasteur., 98:537-556 (1960).

J. Brockwell, Aust. J. Agric. Res., 13:638-649 (1962).

J. Brockwell, in: A Treatise on Dinitrogen Fixation, Sect. 4 (R. W. F. Hardy and A. H. Gibson, eds.), Wiley-Interscience, New York, pp. 277-309 (1977).

J. Brockwell, W. F. Dudman, A. H. Gibson, F. W. Hely, and A. C. Robinson, Trans. 9th Int. Congr. Soil Sci., 2:103-114 (1968).

J. Brockwell, R. R. Gault, D. L. Chase, F. W. Hely, M. Zorin, and E. J. Corbin, Aust. J. Agric. Res., 31:47-60 (1980).

J. Brockwell and R. D. B. Whalley, Aust. J. Sci., 24:458-459 (1962).

J. C. Burton, in: Microbial Technology (H. J. Peppler, ed.), Reinhold Publishing, New York, pp. 1-33 (1967).

J. C. Burton and R. L. Curley, Agron. J., 57:379-381 (1965).

H. D. L. Corby, in: Symbiotic Nitrogen Fixation in Plants (P. S. Nutman, ed.), Cambridge University Press, London, pp. 169-173 (1976).

R. A. Date, Expanded Program of Technical Assistance. FAO Rep. No. 2012, Food and Agricultural Organization, Rome (1965).

R. A. Date, Trans. Int. Congr. Soil Sci. 9th, Vol. 2, pp. 75-83 (1968).

R. A. Date, J. Austral. Inst. Agric. Sci., 35:27-37 (1969).

R. A. Date, Plant Soil, 32:703-725 (1970).

R. A. Date, J. Appl. Bact., 35:379-387 (1972).

R. A. Date, C. Batthyany, and C. Jaureche, Proc. 9th Int. Grassl. Congr., 1:263-269 (1965).

R. A. Date and J. M. Vincent, Aust. J. Exptl. Agric. Anim. Husb., 2:5-7 (1962).

R. A. Date and A. M. Decker, Crops Soils, 14(9):22 (1962).

C. C. Deschodt and B. W. Strijdom, Phytophylactica, 8:1-6 (1976).

M. Engin and J. I. Sprent, New Phytol., 72:117-126 (1973).

R. G. Fawcett and N. Collis-George, Aust. J. Expt. Agric. Anim. Husb., 7:162-167 (1967).

R. A. Fisher and F. Yates, Statistical Tables for Biological, Agricultural, and Medical Research, Hafner, New York (1963).

E. B. Fred, I. L. Baldwin, and E. McCoy, Root Nodule Bacteria and Leguminous Plants, Univ. ov Wisconsin Press, Madison, Wisconsin (1932).

A. H. Gibson, Aust. J. Biol. Sci., 16:28-42 (1963).

A. H. Gibson and P. S. Nutman, Ann. Bot., 24:420-433 (1960).

P. H. Graham and C. A. Parker, Plant Soil, 20:383-396 (1964).

C. Gunning and D. C. Jordan, Can. J. Agric. Sci., 34:225-233 (1954).

C. N. Hale, Proc. 6th Aust. Legume Nodulation Conf., Perth, W. A., pp. 47-49 (1979).

E. Hamatova and H. Vintikova, Inst. Sci. Tech. Info., Prague, 12: 265-280 (1973).

A. Hastings, R. M. Greenwood, and M. H. Proctor, NZ Dept. of Sci. and Ind. Res. Info. Series, No. 58 (1966).

R. A. Hedlin and J. D. Newton, Can. J. Res., Sect. C., 26:174-187 (1948).

A. H. Hendrickson, U.S. Patent, No. 2298561 (1942).

A. E. Hiltbold and D. L. Thurlow, Highlights Agric. Res., 25:12-13 (1978).

V. Iswaran, Madras Agric. J., 59:52-53 (1972).

K. P. John, J. Rubber Res. Inst. Malaya, 19:173-175 (1966).

D. C. Jordan and W. H. Coulter, Can. J. Microbiol., 11:709-720 (1965).

J. Leiderman, Rev. Indus. Agric. Tucuman, 48:51-58 (1971).

L. T. Leonard, J. Bacteriol., 45:523-525 (1943).

J. F. Loneragan, D. Meyer, R. G. Fawcett, and A. J. Anderson, J. Aust. Inst. Agric. Sci., 21:264-265 (1955).

W. L. Lowther, Proc. N.Z. Grassld. Assoc., 38:175-181 (1977).

K. C. Marshall, J. Aust. Inst. Agric. Sci., 22:137-140 (1956).

D. N. Munns, Plant Soil, 28:246-257 (1968).

F. H. Newbould, Sci. Agric., 31:463-469 (1951).

D. O. Norris, J. Exptl. Agric., 31:255-258 (1963).

D. O. Norris, Aust. J. Expt. Agric. Anim. Husb., 11:282-289 (1971).

D. O. Norris, Aust. J. Expt. Agric. Anim. Husb., 11:677-683 (1971).

D. O. Norris, Aust. J. Expt. Agric. Anim. Husb., 12:152-158 (1972).

D. O. Norris, Aust. J. Expt. Agric. Anim. Husb., 13:98-101 (1973).

D. O. Norris, Aust. J. Expt. Agric. Anim. Husb., 13:700-704 (1973).

D. O. Norris and R. A. Date, Legume Bacteriology, CAB Bulletin 51, pp. 134-174 (1976).

C. G. O. Oghoghorie and J. S. Pate, Plant Soil, Special Volume, pp. 185-202 (1971).

H. Philpotts, J. Appl. Bacteriol., 41:277-281 (1976).

H. F. Purchase and J. M. Vincent, Proc. Linn. Soc., New South Wales, 74:227-236 (1949).

R. J. Roughley, J. Appl. Bacteriol., 31:259-265 (1968).

R. J. Roughley, Plant Soil, 32:675-701 (1970).

E. D. Schall, L. C. Shenberger, and A. Swope, Inspection Rep. 106, Purdue Univ., West Layfayette, Indiana (1975).

J. Schiffmann and Y. Alper, Expt. Agric., 4:203-208 (1968).

W. T. Scudder, Proc. Soil Crop Sci. Soc. Florida, 34:79-82 (1974).

M. T. Sherwood, J. Gen. Microbiol., 71:351-358 (1972).

W. A. Shipton and C. A. Parker, Aust. J. Expt. Agric. Anim. Husb.,
    7:259-262 (1967).

M. Sicardi di Mallorca and C. Labandera, Rev. Lat.-Am. Microbiol.,
    20:153-160 (1978).

F. A. Skinner and R. J. Roughley, Annu. Rep. Rothamsted Exptl. Stat.
    Part 1, pp. 86-92 (1969).

F. A. Skinner, R. J. Roughley, and M. R. Chandler, J. Appl. Bac-
    teriol., 43:287-297 (1977).

J. F. T. Spencer and J. D. Newton, Can. J. Bot., 31:253-264 (1953).

J. I. Sprent, New Phytol., 70:9-17 (1971).

J. I. Sprent, New Phytol., 71:603-611 (1972).

M. Stein, M. de Mallorca, and P. M. Williams, Acta Cientifica
    Venezolana, 31:374-375 (1980).

B. W. Strijdom and C. C. Deschodt, in: Symbiotic Nitrogen Fixation
    in Plants (P. S. Nutman, ed.), Cambridge University Press,
    London, pp. 151-168 (1975).

N. S. Subba Rao, S. P. Magu, and K. S. B. Sarma, Arch. Microbiol.,
    77:96-98 (1971).

G. G. Taylor and J. M. Lloyd, Proc. N. Z. Grassland Assoc., 30:154-
    163 (1968).

J. A. Thompson, The Function and Powers of/and Standards Set by the
    Australian Inoculants Research and Control Service (A.I.R.C.S.),
    A Publicity Release, A.I.R.C.S., Gosford, N.S. W., Australia
    (1976).

D. A. van Schreven, in: Nutrition of the Legumes (E. G. Hallsworth,
    ed.), Butterworths Scientific Publications, London, pp. 328-
    333 (1958).

D. A. van Schreven, Van Zee Tot Land, Zwolle, 36:88-108 (1963).

D. A. van Schreven, Plant Soil, 32:113-130 (1970).

D. A. van Schreven, D. Otzen, and D. J. Lindenbergh, Antonie van
    Leeuwenhoek J. Microbiol. Serol., 20:33-57 (1954).

G. Vest, D. F. Weber, and C. Sloger, in: Soybeans: Improvement,
    Production, and Uses (B. E. Caldwell, ed.), Am. Soc. of Agron.,
    Madison, Wisconsin, pp. 353-390 (1973).

J. M. Vincent, Proc. 7th Intl. Grassland Congr., pp. 179-189 (1956).

J. M. Vincent, in: Nutrition of the Legumes (E. G. Hallsworth, ed.),
    Butterworths Scientific Publications, London, pp. 108-123
    (1958).

J. M. Vincent, in: Soil Nitrogen (W. V. Bartholomew and F. E. Clark,
    eds.), Am. Soc. Agron., Madison, Wisconsin, pp. 384-435 (1965).

J. M. Vincent, J. A. Thompson, and K. O. Donovan, Aust. J. Agric.
    Res., 13:258-270 (1962).

J. Vintika and H. Vintikova, Za. Soc. Ch. Nauka, 7:349-360 (1958).

J. A. Voelcker, J. Roy. Agric. Soc. Ser., 3, 7:253-264 (1896).

R. W. Weaver and L. R. Fredrick, Plant Soil., 36:219-222 (1972).

P. M. Williams and M. S. de Mallorca, Acta Cientifica Venezolana,
    31:27-29 (1980).

J. K. Wilson, J. Am. Soc. Agron., 23:670-674 (1931).

J. K. Wilson, J. Am. Soc. Agron., 34:460-471 (1942).

T. Wrobel and J. M. Ziemiecka, Roczniki Nauk Rolniczych, 82-A-1: 201-209 (1960).

ASSOCIATIVE N_2 FIXATION WITH Azospirillum

David H. Hubbell and Murray H. Gaskins

Department of Soil Science
and
Department of Agronomy
University of Florida
Gainesville, Florida 32611. U.S.A.

INTRODUCTION

The association of specific microorganisms with plant roots is an ancient observation. The most obvious examples are pathogenic in nature, where the effects of the microorganism on the plant are usually plainly visible as damage to root tissue, which is also often devastating to the health of the host plant. The study of this general category of root-microbe associations is encompassed in a separate scientific discipline, that of plant pathology. Unique to these associations is the specificity that is involved. For example, it is well established that any plant is not uniformly resistant or susceptible to microorganisms in general. A microbe which destroys one plant species may be without effect on a different species. The recorded number of variations on this theme is without limit. The generalization can then be made that plants of different kinds may be affected by (interact with) specific microbes in the soil environment. The possible reasons for this specificity are incompletely known but are certainly based on the genetics of the organisms involved.

There are also many root-microbe associations that are likewise readily visible but which are symbiotic in nature, and these associations usually provide a distinct advantage to the plant in terms of enhancing its ability to survive and grow in an adverse environment. Notable examples of such associations are the familiar nitrogen-fixing associations consisting of leguminous plants with the bacterium Rhizobium and the association of various nonlegumes with actinomycetes of the genus Frankia. Despite the marked dissimilarity in the organisms involved, these associations have a number of characteristics in common. For instance, the mechanism of root infection in the two systems is remarkably similar. More notable is the fact that, in both cases, the infected root systems are induced to form unique morphological structures, the nodules, which are the sites of symbiotic activity. These structures are readily visible to the eye and are therefore a certain indication of the presence of a root-microbe association. A less obvious manifestation of the presence of such symbiotic associations is the observation of plants growing unusually well in an adverse (low fertility) soil environment.

Within the last decade, evidence has been rapidly accumulating that testifies to the presence of yet another type of root-microbe interaction which may have profound effects on plant growth. We are referring here to the type of interaction now known in the literature as "associative" systems. Such systems, which may involve any of several bacterial species, are best typified by the grass-Azospirillum nitrogen-fixing system first reported by Dobereiner and Day (1975). This type of system is characterized by the much more subtle nature of its establishment and functioning. Under field conditions, there is no readily discernible alteration of root morphology (nodules) in response to microbial infection, nor is the beneficial effect on plant growth sufficiently obvious in all cases to indicate the presence of a desirable microbial population in the rhizosphere. These systems have been recently termed "cryptic" associations since they are generally macroscopically undetectable or "hidden" (Hubbell and Gaskins, 1980).

In order to understand fully the nature and implications of these "cryptic" root-bacteria associations, it is important to be aware of a general concept of soil-plant-microbe relationships that has been defined as the "rhizosphere" (Rovira, 1969). This term refers to the idea that substances derived from the roots of different plants have profound and, to varying degree, specific effects in determining the numbers and kinds of microbes which predominate in the soil immediately adjacent to living roots. The rhizosphere is therefore defined as the portion of the soil within the zone of influence of plant roots. In this zone, subtle root-microbe interactions may occur frequently and perhaps always (Dommergues and Krupa, 1977).

Recent intensive research efforts on many aspects of the grass-Azospirillum association have resulted in an accumulation of information which justifies our detailed discussion of this particular "associative" system. Only time and much additional research will reveal whether this particular example is most typical of associative systems in general. The recorded number and diversity of these systems are increasing with time. We will present observations on the formation and functioning of such systems, identify gaps in our cur-

rent understanding, and speculate on the future course and significance of research on this subject. The grass-Azospirillum system
is perhaps best characterized and possibly most representative of
"cryptic" root-microbe associations. For that reason, we have chosen
to dwell primarily on the details of this system as an example of
what is possible in such associations. Because the literature on
this subject is quite extensive, we have not attempted to present
a comprehensive review. Instead, a limited number of references
that emphasize or illustrate points discussed herein are noted in
the text. We extend apologies to uncited authors for the fact that
much important work is not mentioned specifically.

HISTORICAL BACKGROUND

     The excellent review of N_2-fixation in grasses by Neyra and
Dobereiner (1977) documents the widespread distribution of N_2-fix-
ing microorganisms in soil and the early suggestions that they could
contribute significantly to the supply of soil nitrogen. Research
on the general phenomenon of associative N_2 fixation in grasses received great impetus when Dobereiner (1961) showed that N_2-fixing
bacteria of the genus Beijerinckia are selectively stimulated in the
rhizosphere of sugarcane. This may explain the observations that,
in many parts of the world, this crop has been grown in mono-culture
for over 100 years without the addition of fertilizer nitrogen and
fields frequently fail to respond to fertilization even after addition of P and K. Subsequently, Dobereiner (1970) showed a specific
association of Azotobacter paspali with certain ecotypes of Paspalum
notatum. Dobereiner and Day (1976) later reported a third system
made up of the important tropical forage grass Digitaria decumbens
with the N_2-fixing bacterium Spirillum. This bacterium, now classi-
fied as Azospirillum, has been intensively studied since that time
and therefore serves as a representative example of associative N_2-
fixing systems in this presentation. It should be emphasized once
again, however, that numerous soil bacteria that are capable of fixing nitrogen are known to proliferate in the rhizosphere of different plants, from which they are readily isolated. This includes such
genera as Bacillus, Enterobacter, Pseudomonas, Klebsiella, Erwinia,
Azotobacter, Beijerinckia, and others. The list of agronomically
important forage and grain crops shown to harbor such associations
grows longer and now includes Brachiaria, Digitaria, Panicum,
Pennisetum, Cynodon, and Setaria as well as corn, sorghum, rice, and
wheat.

THE BACTERIAL COMPONENT

     The microorganism associating with Digitaria was first tenta-
tively identified as Spirillum lipoferum. Subsequent detailed taxonomic study has established that these organisms are sufficiently dis-

tinct to merit status as a new genus, Azospirillum. The genus contains two species. These are differentiated mainly on the basis of carbohydrate metabolism, A. brasilense being oxidative, whereas A. lipoferum is fermentative.

Azospirillum is routinely isolated from soil and root samples using a semisolid N-free mineral medium containing potassium malate (Dobereiner and Day, 1976). By the second day after inoculation, a fine, dense, white pellicle forms 1 to 2 mm below the surface. Cells from this layer may be used as inoculum for a second enrichment on the malate medium, and this enrichment is followed by streaking on the malate medium supplemented with yeast extract. The latter component is required because growth of Azospirillum will be inhibited by the presence of oxygen if combined nitrogen is not supplied. After 5 days, the organism will appear as small, dry, white or pinkish, raised, round or irregular dense colonies.

Azospirillum has been isolated from soil and root samples in Europe, Africa, North and South America, and other locations. The organism appears to associate preferentially with tropical grasses. However, recent research indicates that there are numerous exceptions to the rule. The organism is generally found wherever it is sought. It may ultimately be concluded that Azospirillum is quite promiscuous in the kinds of plants with which it will associate. It is possible that the apparent preference for tropical grasses is more quantitative than qualitative.

Numerous studies confirm that there are two main physiological types within the genus Azospirillum. The first of these groups, designated Azospirillum brasilense, possesses primarily an oxidative metabolism. Sugars are not used well but organic acids such as malic, succinic, and fumaric acids are readily utilized for both growth and N_2-fixation. The second group, designated Azospirillum lipoferum, possesses a greater ability to ferment certain sugars with the production of acid. Both groups are capable of utilizing a wide variety of compounds as sole carbon sources for growth, the primary distinction among the two groups lying in the ability of A. lipoferum types to produce acid. Amino acids and amines are also used as sole carbon and nitrogen sources for growth of both groups.

Azospirillum is able to mediate a wide variety of nitrogen transformations. Although most attention has been given to its N_2-fixing capability, the ability of the bacterium to carry out several other reactions of the nitrogen cycle is also well documented. The organism is capable of using ammonia, nitrate, and a variety of organic nitrogenous compounds for aerobic growth. Ammonia and nitrate are utilized equally well in aerobic culture.

Growth of Azospirillum on atmospheric dinitrogen occurs most rapidly at substantially reduced partial pressures of oxygen. The optimal pO_2 for N_2 fixation ranges from 0.005 to 0.007 atm. At this optimum, A. lipoferum assimilates 8 to 10 mg of nitrogen per gram of carbon substrate consumed. The necessity for greatly reduced pO_2 for N_2 fixation is no doubt a reflection of the sensitivity of nitrogenase to oxygen. The formation of "pellicles" (well-defined zones of growth below the surface of the medium) by Azospirillum when grown in semi-solid N-free agar deeps, considered by some to be diagnostic of the presence of this organism, reflects the ability of the organisms to grow at a depth where the pO_2 is favorable. It is to be stressed, however, that many other microaerophilic non-N_2-fixing bacteria form similar pellicles in culture media contained in tubes. Therefore, the formation of pellicles in semi-solid N-free media cannot be used as a sole criterion for the presence of Azospirillum-like organisms, nor can pellicle formation combined with positive acetylene reduction be so used because other diazotrophs can exhibit similar growth patterns.

Azospirillum behaves as an aerobe when grown in the presence of elevated levels of combined nitrogen. In N-free media, it is microaerophilic. Low levels of combined nitrogen, such as yeast extract or casamino acids, may stimulate the onset of N_2 fixation in "N-free" media. The organism grows well on some sugars (glucose, sucrose, galactose, lactose, or mannitol) when the culture is supplied with low levels of combined nitrogen or malic acid.

Another trait observed among several strains of Azospirillum (13t and Sp7) is the ability to carry out denitrification. The same conditions that favor nitrogenase activity (i.e., low pO_2 or anaerobiosis) also favor denitrification. The ecological and agronomic significance of this trait has yet to be fully investigated.

In vitro studies show that many species and strains of rhizosphere bacteria achieve high rates of N_2 fixation under optimum conditions. Because the capability is a common one, it is not satisfactory as a selection criterion. Capacity to survive and multiply in the rhizosphere is a much more important performance characteristic than is the relative capability for N_2 fixation by pure cultures.

BIOLOGY OF THE ASSOCIATION

Characteristics that determine the capacity to grow in the rhizosphere may be summarized in three categories: ability to withstand the physical and chemical conditions prevailing in the soil (survival capability), ability to grow well using only the carbon and mineral supplies available in the rhizosphere (metabolic capability), and ability to compete successfully with other organisms present in the rhizosphere for the available supply of growth-supporting substrates (competitive capability). Some of the major problems that must be studied if associative N_2-fixing bacteria are to be exploited to benefit crop production will be noted under these categories.

Intense interest in associative N_2-fixing bacteria developed after some early investigations indicated that they grew and fixed nitrogen inside living plant cells. However, this contention has not been confirmed by subsequent investigations. There is no clear evidence that Azospirillum infects living plant cells. Instead, compelling evidence has accumulated to indicate that the bacterial cells infect moribund or dead cells of the root cortex (Umali-Garcia et al., 1978). This implies that Azospirillum may not fix appreciable nitrogen within the root.

Various Azospirillum strains can utilize pectic substances as energy substrates (Tien et al., 1981). We do not know the extent to which they do so under natural conditions. There is evidence that Azospirillum cells grow to a limited extent in the intercellular spaces of roots. Colonies often develop at sites where senescing root hairs provide entry points. It is not clear if the Azospirillum cells at such points are utilizing as energy substrates the pectic substances of the lamellar layer or the cytoplasmic components of the lysed epidermal cell.

The mucigel layer found on growing root tips is colonized by Azospirillum in gnotobiotic culture (Umali-Garcia et al., 1980), and while this has not been demonstrated under natural conditions, it probably occurs. As noted previously Azospirillum strains have the capability to hydrolyze pectic substances, a major constituent of mucigel. Because of its viscosity, the mucigel is certain to act as a barrier to gas diffusion and therefore might provide a microaerophilic environment conducive to nitrogenase activity in Azospirillum. Further, nitrogenous compounds such as amino acids are scarce or absent, because cytoplasmic compounds such as those released by cell lysis are not present. Thus, in the mucigel, a suitable carbon substrate is at hand in an environment in which the levels of organic nitrogen and oxygen are low, and nitrogenase activity should not be repressed.

Microbial growth occurs at the region of maturation (Umali-Garcia et al., 1980), a point where a supply of carbon substrates is produced on the growing roots by lysis of senescent root hairs and other epidermal cells. In contrast to the mucigel, which is an excreted product, the constituents of the previously living cells include nitrogenous components such as amino acids, and therefore bacteria growing at these sites should be able to grow without fixing a new supply of nitrogen. Further, because the viscous mucigel is not produced at this location, there may be more oxygen at the site. Thus, the higher levels of nitrogenous material and of oxygen may act to suppress N_2 fixation. Certainly, Azospirillum is fully

capable of utilizing the organic acids, amino acids, and monosac-
charides produced upon the release of cytoplasmic components, when
the senescent plant cells are degraded. However, this capability
is shared by competing organisms.

Numerous approaches have been used to study conditions that
support growth of Azospirillum in the rhizosphere. However, these
efforts have not identified specific physical, chemical, or biochemical circumstances that cause these bacteria to proliferate
selectively or preferentially. Even when favorable plant responses
are noted following Azospirillum inoculation, bacterial numbers
nevertheless decline. However, various studies suggest that specific, beneficial interactions between Azospirillum strains and
plant cultivars may occur and that these are genotype-specific, as
are those of symbiotic associations. The clear demonstration of the
biochemical basis for this type of relationship between plants and
"associative" bacteria most certainly would be a pivotal development, since the conditions necessary for the interactions between
the two genotypes then could be investigated and defined. At this
point, however, we cannot identify unequivocally any plant or plant
product that selectively promotes growth of Azospirillum, nor have
we demonstrated beyond doubt that inoculation with Azospirillum results in a stimulation of plant growth not achievable by inoculation with other bacteria.

Azospirillum strains survive in the rhizosphere following soil
inoculation, but invariably the populations stabilize at quite low
levels. In many greenhouse experiments, with a variety of soils,
plants, and growth conditions, we have found this to be the rule.
No physical or chemical treatments have yet been found which bring
about an increase in these numbers. With appropriate allowances for
difficulties in counting and resultant possible underestimation of
bacterial numbers, these observations nevertheless indicate a major
obstacle to agronomic usefulness of the bacteria. The present evidence seems to show quite clearly that if rates of N₂ fixation are
to be increased substantially, methods must be found to increase
selectively the rhizosphere population of bacterial strains that
fix nitrogen. The level of energy substrates in the soil can be increased by the addition of readily metabolized carbon sources, but
such methods simply promote uncontrolled growth of many organisms
and do not achieve selective proliferation of desirable bacteria.
This will be discussed further in the following section.

Upon prolonged incubation on solid media, Azospirillum produces
thick-walled, cyst-like ovoid structures that are more resistant to
desiccation than are vegetative cells. Cyst production may thus provide a mechanism for survival of Azospirillum in the rhizosphere during unfavorable environmental conditions.

Developing plant roots, as they grow, disturb the existing
equilibrium among soil microorganisms. New roots may alter the pH,

pO$_2$, moisture levels, and many other environmental conditions that selectively affect growth rates of soil organisms. Most studies aimed at manipulating soil populations have been concerned with the control of phytopathogenic species, and some rules governing cultural techniques to suppress the development of pathogens are well recognized. Such techniques normally fail to provide absolute protection, but nevertheless highly destructive plant diseases such as bacterial wilts, cankers, and blights can be suppressed by manipulations of such soil conditions as pH and moisture levels. Whether the reverse can be achieved, the selective propagation in the soil of desired bacteria, may await discovery of combinations among soil conditions and plant genotypes that cause Azospirillum or other beneficial organisms to multiply selectively. This would require a technique by which competing organisms would be placed at sufficient disadvantage so that their utilization of available energy substrates would be severely restricted. As we have noted repeatedly, Azospirillum and other beneficial N$_2$-fixing bacteria grow well in many environments when their competitors are suppressed. If this cannot be arranged under field conditions, we must hope that other methods may be found to promote their growth and restrict the growth of their competitors. This may be possible only if plant-bacterial combinations are discovered in which the level of interaction between the two approaches is characteristic of symbiotic combinations. It is possible that such combinations exist in nature. Data such as those reported from studies of field plots in Israel suggest a high level of plant-bacterial interactions, which are sustained through the lifetime of the host plants and are of substantial benefit to them. Whether these responses are similar in kind, and merely greater in magnitude for some unknown reason, than those discussed in earlier reports remain to be determined. It seems improbable that responses to bacterially produced growth hormones could account for the large differences in yield that have been reported. Because the advanced farming methods common in Israel normally include liberal use of mineral fertilizers, it seems unlikely also that bacterial treatments would affect crop yields substantially by promoting cation absorption, an effect more likely to be significant when soluble fertilizers are applied sparingly or not at all. Thus, the Israeli experiments suggest that N$_2$ fixation by the inoculant bacteria may have occurred, and may have brought about the increased yields reported.

A recurring question, which has not yet been satisfactorily answered, is whether energy substrates in the rhizosphere are sufficient to support significant levels of growth and N$_2$ fixation by associative bacteria. The production of "exudates" by living roots has been studied by various methods, all of which have certain difficulties (Rovira et al., 1979). Tracer experiments with ^{14}C-labeled compounds have been used by us and others to determine rates at which plant metabolites become available to root-associated microorganisms. In our experiments, we find that production of lysate or

exudate is directly proportional to plant growth rates. Under
growth-limiting conditions, exudate production is restricted. These
methods provide highly useful information, but they are not easily
adapted to mass screening. When plants are grown in nutrient solutions, it is a relatively easy matter to analyze the solutions and
report the concentrations of various substrates left there after
the roots are removed. Such studies invariably show that measurable
but rather small amounts of organic acids, carbohydrates, amino
acids, and various other plant metabolites are present. The quantities reported may be lower than those actually produced because
microorganisms, when present in the solutions, readily absorb and
utilize most of these substances as growth substrates. The microorganisms themselves can be analyzed of course, but the resulting
data cannot then separately identify those substances produced by
the plants. To circumvent this difficulty, plants may be grown
gnotobiotically (Hale et al., 1978). This, however, does not resolve the issue. We and others have noted that plants grow less
well in gnotobiotic than in normal culture, and that total carbon
flux from the root system is reduced by gnotobiotic culture conditions. Pulse-labeling experiments show that as much as 10 to 20% of
the carbon fixed by leaves may be exported in one or another form
from the roots, but our results indicate that the portion of this
that is readily metabolizable by Azospirillum or similar bacteria is
rather small. We cannot form valid conclusions simply by comparing
the normal and gnotobiotic systems, however, because of the physiological interactions between the bacteria and the roots. Further
studies of this type are very badly needed to better understand the
metabolic coupling that occurs between bacteria and roots and to
determine whether means can be developed to support use by N$_2$-fixing
bacteria of a larger share of the available energy substrates. A
substantial amount of information has been collected from in vitro
studies, and it seems doubtful that further repetition of such experiments will be of much use. In vivo studies, although much more
difficult and time consuming, eventually will provide the basis for
further development of associative N$_2$-fixing systems.

A large volume of Russian literature has been published on soil
inoculation with Azotobacter and other bacteria, fungi, and actinomycetes. Numerous instances of beneficial crop responses were attributed to N$_2$ fixation by the bacteria. This literature was critically reviewed, and it was concluded that many of these claims were
exaggerated or unfounded (Mishustin and Naumova, 1962). It was
pointed out that all types of microorganisms could produce similar
beneficial plant response, thus suggesting a common mechanism which
is apparently widespread in the microbial kingdom.

Subsequent work on this controversial subject has been reviewed
by Brown (1972). Some of these studies presented evidence that plant-
growth hormones produced by the microorganisms could be responsible
for the observed plant response. This was confirmed with Azotobacter

inoculated on various grasses and other crops (Brown, 1976). In a more recent study, numerous genera of soil bacteria were isolated from the wheat rhizosphere and shown to produce plant growth-regulating compounds. The organisms were most abundant on roots of older plants. Inoculation of plants with the various organisms produced plant effects similar to those obtained by adding only the pure compounds (Brown, 1974).

Tien et al. (1979) demonstrated production by an _Azospirillum_ strain of auxin (IAA), cytokinins, and substances tentatively identified as gibberellins. These studies were with pure cultures, and there is as yet no definite proof that the bacterium produces these substances when growing in the rhizosphere. However, we have found that _Azospirillum_ inoculation increases the cytokinin content of xylem exudate from decapitated plants. The apparent responses of plants to growth-stimulating hormones produced by root-associated bacteria emphasize the need to exercise caution in the interpretation of yield increases in response to bacterial inoculation. There is no question that bacterially produced hormones can affect plant growth. Some of the mechanisms that may be involved will be mentioned briefly.

The critical role of auxins in many growth processes is well known. However, it is not at all clear that auxins produced by rhizosphere bacteria are significant. In many, but perhaps not in all instances, plants normally produce auxin in excess of their needs, and regulatory processes involve enzymatic degradation of excess amounts, as well as synthesis, at sites where auxin regulates physiological activity. Nevertheless, the premise of Frits Went, stated 50 years ago, is now fully accepted: without auxin, there is no growth. Because the mechanism of auxin degradation in plants is highly efficient, experiments to study auxin effects often are performed with analogs, such as 2,4-D, which are not subject to degradation by enzymes that deactivate auxin. Studies with such materials occasionally show yield improvements, suggesting that at least in some circumstances auxin levels above those found in normal plants can have positive effects.

Much recent research involves study of auxin regulation of cation uptake and cell growth. However, some recent studies clearly implicated auxins in cell differentiation processes as well. Auxin enhances release of various carbohydrates from cell walls, and it is possible that the production of auxin by rhizosphere bacteria leads to release from roots of carbohydrates that can be used as energy substrates by the bacteria.

Cytokinins are known to be involved in the regulation of several vital processes. Apparently, cell division cannot occur in the absence of cytokinins, because conditions that prevent their synthesis also block reproduction of cultured plant cells. Cytokinins

strongly inhibit senescence of leaf tissue, and this physiological
effect is the basis of several bioassays. Cytokinins are believed
to be synthesized primarily in the roots of plants and translocated
in the xylem stream to sites of activity. Many bacteria synthesize
cytokinins, but the significance of this activity is difficult to
establish. A method for distinguishing cytokinins of bacterial
origin from those produced by plant tissue would permit this to be
determined. Since it is clear that cytokinin transported in the
xylem stream to shoots exercises strong control over developmental
processes, there is a great need to identify that portion of total
cytokinin production in the root system that is attributable to bacteria and other microorganisms growing in or on the roots.

The gibberellins occur in several forms, which share the basic
skeletal structure but differ in end-groups that determine biological and chemical activity. The various chemical structures are
difficult to separate and identify, and because of this, the number
of naturally occurring gibberellins has not been determined. As
with various other biologically important materials, extraction and
assay procedures may produce artifacts, and therefore conclusions
about "natural" forms must be drawn with care.

Although gibberellins were discovered as fungal products, most
research has been focused on their occurrence in plants and their
significance as regulators of plant growth and differentiation. The
most obvious and striking effects of gibberellins are those in which
cell elongation is involved, but it is clear that many other highly
important growth functions are controlled by gibberellins. Reports
about gibberellin production by bacteria are less numerous than reports about auxins and cytokinins, but it is clear that many bacterial species have the capability. As noted for the materials mentioned above, distinguishing between plant products and products absorbed by the plant from bacteria or other microorganisms is often
difficult. Therefore, in studying natural plant-bacterial associations, it is not always possible to identify beyond doubt the origin
of gibberellins found in the plants.

The regulatory effects are particularly conspicuous at the time
of seed germination. The control of amylase development and the
subsequent effects of amylolytic enzymes in mobilization of cotyledon-
ary reserves to support development of the germinating embryo have
been studied extensively. This control mechanism has been demonstrated in studies involving seed germination at low temperatures,
which suppress reserve mobilization. Under such circumstances, very
small amounts of gibberellins often induce substantial changes in
rates of seed germination. The significance of gibberellins produced by bacteria under these circumstances has not been determined.
However, because bacterial colonies develop rapidly at the surface
of germinating seeds as a result of exudation of plant metabolites
from the cotyledons, it is virtually certain that some if not all

of the rapidly growing bacteria would produce not only gibberellins
but other growth substances as well.

Still other interactions may be involved when plant growth is
improved by bacterial inoculants. Since the soil microflora normally contains both beneficial and harmful bacterial species, it is
clear that any influence that causes one of these groups to prosper
may affect plant growth. There is evidence that soil inoculation
with bacteria not directly beneficial to plants may nevertheless in
some cases produce beneficial effects on yield, and these effects
result from suppression of harmful microorganisms. The extent to
which this mechanism may be used in agriculture remains to be determined. A method based on this principle for controlling soil-
borne pathogens most certainly would have many attractive features.

A further effect of soil bacteria is to solubilize minerals and
thereby increase their availability to plant roots. Iron and phosphorus may be made available in adequate quantities under conditions
that otherwise would not permit satisfactory plant growth. Because
many soils otherwise suited for agriculture lack sufficient iron or
phosphorus for good plant growth, this mechanism deserves careful
study. Plant species vary widely in their tolerance to nutrient
deficiencies, and within crop species some cultivars grow much more
efficiently than others under nutrient-limited conditions. The
mechanisms that confer these characteristics have not been defined
in many cases. It is clear, however, that rhizosphere bacteria are
involved in some instances.

THE PLANT COMPONENT

Perhaps one of the most significant and encouraging findings in
relation to associative N_2 fixation is that there appear to be
varietal differences in response to inoculation with N_2-fixing bacteria. Varieties responding more favorably than others have been
reported for several grasses, including sorghum, wheat, sugarcane,
and others (Vose and Ruschel, 1981).

Von Bulow and Dobereiner (1975) reported significant differences among S_1 lines of maize in field trails in Brazil. The criterion for enhanced N_2 fixation potential in this study was elevated
levels of acetylene reduction determined from preincubated washed
roots. Their best lines in this study showed nitrogenase (acetylene
reduction) activity several times higher than activity found in other
lines. Their findings led the authors to believe that the potential
nitrogenase activity in maize roots approached that of soybeans.
Later studies have shown, however, that the assay technique (in
vitro acetylene reduction assays conducted with excised roots) seriously
overestimated N_2 fixation (Gaskins and Carter, 1975; Van Berkum and

Bohlool, 1980). Nevertheless, it is significant that there was a marked difference in acetylene-reduction activity among the different lines, which was attributed to "Spirillum sp." in the rhizosphere.

A similar genotype response for enhanced N₂ fixation was reported for a particular variety of wheat by Neal and Larson (1976). In that work, a N₂-fixing Bacillus was consistently recovered from the rhizosphere of a chromosome-substitution line of spring wheat, whereas none of the bacilli was recovered from either of the two parent cultivars (Rescue and Cadet). In this case, effects on acetylene-reduction rates and plant yields were not studied. The report, however, indicates that it may be possible to alter the rhizosphere through genetic manipulations of the host plant in order to favor certain diazotrophic bacteria.

Bouton et al. (1979) studied the response of Pennisetum americanum (pearl millet) to inoculation with A. brasilense, and found that one hybrid cultivar (Gahi 3) produced significantly higher dry-weight yields in response to inoculation in the field. Inoculated plants produced 32% more dry weight and contained 37% more total nitrogen than controls. However, acetylene-reduction values were low and not correlated with yield increases. Similar results were obtained with greenhouse plants. The authors concluded that genotypes of pearl millet differ in response to bacterial inoculation, but they were unable to determine whether N₂ fixation by the inoculant organisms contributed to the increased plant growth. They suggested that further study would be necessary to identify the response conclusively.

Still more evidence for varietal differences comes from the work of Ruschel and Ruschel (1977) on the nitrogenase activity of the sugarcane rhizosphere. Two varieties (NA 56-62 and CB 46-47) gave significantly higher rates of acetylene reduction, and the activity was greatest in "undisturbed" systems inoculated at decreased oxygen tensions (pO₂ of 0.2 atm). A high correlation between plant N concentration and acetylene reduction in five of the six varieties was observed for plants grown in the same soil types. It was noted that mature stalks of varieties CB 46-47 and NA 56-62 were rich in sucrose whereas stalks of variety CB 41-76, which showed almost no acetylene reduction, contained little sucrose. A possible relationship between sucrose content and nitrogenase activity was suggested.

Several conclusions are suggested by these studies. It is increasingly apparent that there are definite genetic effects of the host plant on nitrogenase activity in the rhizosphere-rhizoplane. These differences may arise from quantitative or qualitative changes in the nature of root exudates upon which the energy demanding diazotrophs are dependent. If so, and if root exudates can be modified to produce substrates favored by specific diazotrophs, it might

be possible to enhance colonization and N_2 fixation by root-associated bacteria.

Since associative N_2 fixation is largely dependent upon some
type of plant for the energy (carbon) required in the process, a
discussion of this aspect of the association is warranted. Perhaps
the first question should be, "How are these associations between
various combinations of diazotrophs and plants maintained energetically?" No simple answer to this question can adequately explain the myriad of biological phenomena involved in such an association. Perhaps the most obvious explanation for the existence of
these associations is the availability of carbon and energy sources
in the rhizosphere and immediately at the root surface. Historically, a great quantity of literature has accumulated on the so-
called "rhizosphere effect." It is well known that a variety of
metabolic processes are to some degree enhanced in the rhizosphere.
These increases are in most instances a result of the increase in
the availability of growth substrates liberated by the roots (exudates and sloughed cell debris). The fact that N_2 fixation is enhanced in the rhizosphere and still more at the root surface (rhizoplane) is again a reflection of the availability of nutrients. In
view of the high energy demands for biological N_2 fixation (15 to
33 moles of ATP per mole of N_2 fixed), the process is highly subject
to limitations imposed by the supply of energy-yielding substrates.
Thus, factors which influence the rate of photosynthetic activity
maintained by plants, the amount of photosynthate translocated to
the root system, and the quantity of carbon compounds which ultimately become available to the root-associated bacteria are certain
to be rate limiting.

It is possible that plants highly efficient in the production
and translocation to the roots of photosynthetic products might provide better nutrient supplies to root-associated bacteria. By virtue
of several properties not exhibited by C_3 plants, the C_4 plants are
more efficient in photosynthesis and hence primary production. This
higher efficiency might be responsible for the fact that many of the
N_2-fixing plant-bacterial associations reported involve C_4 plants.

While there is little doubt that the C_4 plants fix CO_2 more
efficiently than C_3 plants, this enhanced capability is not always
reflected in plant productivity. Gifford (1974) has reviewed maximum recorded values of various parameters for the rate of CO_2 fixation from the biochemical level through the leaf level to crop growth
rate for C_3 and C_4 species. He suggests that the "large potential
advantage of the C_4 mechanism at the biochemical level is progess-
ively attenuated in moving from the microscopic to the macroscopic
parameters until, at the level of crop growth rate, there is no difference between best examples of the two groups when grown in their
own preferred environments."

R. H. Brown advanced the hypothesis that, in addition to the differences already discussed, the C_4 grasses are more efficient than C_3 plants in utilizing available combined nitrogen. His proposal is based on (a) the higher dry matter production of C_4 species, (b) lower N concentrations in C_4 species, (c) a reduction in C_4 species in N invested in fraction 1 protein (RuDP carboxylase), whereas C_3 species invest up to 50% of their soluble leaf protein in fraction 1, and (d) the mechanism in C_4 grasses for maintaining high concentrations of CO_2 in bundle sheath cells surrounding the RuDP carboxylase enzyme. He further speculated that this greater efficiency of nitrogen use in C_4 plants may give them a competitive advantage in most climates with favorable temperatures, particularly when soil N is limiting. In summary, while C_4 plants have been shown to have capabilities that in some circumstances increase their efficiency, some C_3 plants appear to be equivalent to C_4 plants when grown in their own preferred habitats. It remains to be determined whether the C_4 photosynthetic pathway enhances biological N_2 fixation by associative bacteria.

PROBLEMS AND PROGRESS: PAST, PRESENT, AND FUTURE

The kinds, numbers, and activities of microorganisms associated with plant roots (rhizosphere microorganisms) are determined by three kinds of factors. First, there are soil factors, including structure, texture, water potential, gas phase composition, pH, temperature, nature of organic and inorganic components, and the microbial community. Second, there are climatic factors, including light, air temperature, relative humidity, and CO_2 concentration, that influence the rhizosphere microorganisms through their effects on the physiology of the plant. Finally, there are factors introduced by the roots of growing plants. The physiology of the roots of any given species is the sum of many different genetically determined traits characteristic of that species. The interaction of metabolically active root tissue with biotic and abiotic components of its environment produces a state of continuous change in that environment.

It is axiomatic among microbiologists that "the microorganism does not select the environment - rather, the environment selects the microorganisms." Because of their size and ubiquitous nature, most microorganisms are routinely introduced (inoculated) into new environments by many means. Their ability to survive in their new environment is a function of the ability of the environment to satisfy the growth requirements of the organism.

Now let us consider the present and future agronomic potential of inoculation with Azospirillum or any other associative bacterium. Given that soil, climatic, and plant factors will combine to determine the success of inoculation, it can be inferred that success will

be dependent on our ability to control or regulate these factors
within the limits defined by the microorganisms.

The theoretical possibility for high levels of N_2 fixation by
soil bacteria is clear, and it is this possibility that has sharply
stimulated research on root-associated bacteria in the past decade.
Some research combined with substantial amounts of speculation in
the early 1970's produced many high hopes. These hopes diminished
in the last half of the decade, and many researchers who began
studies with root-associated bacteria subsequently dropped their
work in disappointment. Ironically, rapidly developing expertise
in techniques of genetic manipulation provide many opportunities
for productive research that were not open until very recently.
These will not be discussed in depth here. It is important, however, to emphasize that the techniques and tools for resolution of
the problems and limitations noted herein may be close at hand or
already available. While a great deal of work must be done, there
is good reason for optimism about achieving highly significant advances in this research area. It is not realistic to suppose that
the performance of microorganisms in controlled laboratory experiments can be duplicated under natural conditions. It is quite reasonable, however, to expect that, through effective use of research
techniques now at hand, plant-bacterial interactions can be developed and used effectively to increase crop production.

It is apparent that agricultural soils support associative bacteria generally, but numbers and/or activities are low. Since the
desired organisms are present, it is implied that inoculation is not
necessary. Instead, the environment must be altered to favor the
maintenance of higher populations of microorganisms. It is frequently not possible, or at least not economically feasible, to control climatic or soil factors on a large scale.

Is there, then, any reason to hope that we can, in the future,
alter soil environments in such a way as to favor associative rhizosphere microflora? We believe there is reason for optimism. As
mentioned earlier, these potentially beneficial associations have
been shown to occur under a variety of soil and climatic conditions,
and to involve a variety of plants. They apparently exist in most
and perhaps all agricultural soils. They are therefore limited not
in existence in absolute terms, but in extent of existence. The
question that is perhaps most applicable is not, "How can we inoculate successfully?" but rather, "How can we alter the environment to
increase organisms already present?"

As we mentioned earlier, plants differ in the extent to which
they will "associate" with different microorganisms. In other words,
there is some degree of host specificity demonstrated in the establishment of these systems, which is evident as differences between
species as well as between cultivars within a species. It can be

concluded that something unique in the physiology or metabolism of
the roots of plants encourages or discourages the development of
these associative systems. This is a restatement of the definition
of the "rhizosphere effect." The metabolic characteristics of plant
roots that result in these effects are clearly based in the genetics
of the plant. The genetically determined metabolic characteristics
of the plant determine the nature of the rhizoplane-rhizosphere environment, which selects for or determines the kinds and numbers of
organisms that will survive and establish in it. Advances in the
science of plant breeding and selection present the clear possibility of developing cultivars of desirable agronomic crops with metabolic characteristics, new or enhanced, that favor formation of associative systems. To accomplish this objective with any degree of
speed and efficiency, it is first necessary to identify those metabolic traits of the plants and the bacteria that are critical and to
elucidate the nature of their interaction. It should then be feasible to "engineer" plant cultivars that will, by virtue of unique
metabolic traits, create a unique rhizoplane-rhizosphere environment that will select for a beneficial rhizosphere microflora.

It is possible that N_2 fixation occurs in primitive genotypes
but not derived cultivars. By selecting for maximum response to
nitrogen fertilizers, breeding programs may have selected against
the ability to form N_2-fixing associations. If so, future breeding
programs may benefit from concentrating on primitive genotypes. It
might be interesting to compare plant physiological characteristics
of primitive genotypes and advanced cultivars derived from them.
Such comparisons could identify traits required for establishment
of such associations, thus providing invaluable guidance for future
breeding programs. Quantitative differences in production and translocation of photosynthate to the roots may be important. These factors are known to influence the extent of N_2 fixation in some legumes.
Qualitative and/or quantitative differences in exudates "leaking"
from roots are an important factor in establishment and functioning
of bacteria-root associations. Differences in exudates between genotypes may well be correlated with differences in N_2-fixing activity.
Study of such differences and their significance should be approached
with caution because reports indicate marked differences in results
from laboratory and field studies. The physical environment of the
plant is a critical factor. It might be possible to develop plant
genotypes that "leak" energy substrate sufficient in quality and
quantity to support significant activity of associative N_2-fixing
bacteria. However, as noted above, our experiments in which exudation was increased by experimental manipulation did not produce
highly encouraging results.

Bassham (1977), commenting on the potential for increasing crop
production through plant genetics, discusses many points relevant to
associative N_2 fixation. One of the most noteworthy of these is the
dramatic yield increases exhibited by C_3 plants when grown under con-

ditions of CO_2 enrichment. High CO_2 levels suppress photorespiration and increase the photosynthetic efficiency of C_3 plants. Plants selected for low photorespiration rates might be more efficient and therefore capable of supporting larger populations of N_2-fixing bacteria. That enhanced N_2 fixation can occur as the result of CO_2 enrichment has been well documented in studies with Rhizobium-legume symbioses. Similar studies with associative systems are needed. Since one of the major factors regulating root-associated N_2 fixation may be the supply of photosynthate to the root system, any mechanism that increases the availability of carbon compounds at the root surface may also increase associative N_2 fixation.

Development of free-living diazotrophs not represssed by elevated levels of combined nitrogen might lead to increased N_2 fixation. Schubert and Evans (1976) have reported strains of Rhizobium that have N_2 fixation efficiency as much as 30% above normal, owing to the presence of an uptake hydrogenase that recycles the hydrogen generated by nitrogenase activity. Whether this mechanism can be effective in associations with free-living diazotrophs remains to be investigated. Brill (1975) and his coworkers have identified regulatory mutants of free-living diazotrophs that produce high levels of nitrogenase even when excess combined nitrogen is present. Determining whether such mutant strains can be of agronomic importance will require extensive study of their competitive ability in natural systems, where competing native strains are in abundance.

The occurrence of pectinase in Azospirillum has already been discussed. The occurrence of extracellular enzymes capable of modifying plant cell-wall materials (pectinase, etc.) may be a general phenomenon in many common genera of soil bacteria. These bacteria, many of which are demonstrated N_2 fixers, associate with plants to varying degrees. This implies the presence of a continuous source of carbon for growth and N_2 fixation. The kinds and amounts of carbon compounds released by the roots vary among plant genotypes, and one can reasonably infer that these characteristics are under genetic control. It might be possible to develop genotypes that associate with selected bacteria because of genetically based enhancement of specific biochemical characteristics. This would constitute "genetic engineering" of plant metabolism to obtain plants which, in the course of normal growth, would stimulate preferentially the growth and activity of certain indigenous soil bacteria having the inherent capacity to benefit plant growth. This presupposes a rather thorough knowledge of the mechanisms involved in such associations. Two properties of bacteria already identified as important in this respect are (a) the ability to fix nitrogen and (b) production of plant growth substances. Our incomplete knowledge of the biochemistry of beneficial plant-bacterial interactions may dictate that initial plant breeding and selection studies be conducted empirically. Adverse plant growth conditions, such as minimal soil nutrient status, should be used in these studies, since microbially

induced beneficial effects on plant growth should be made obvious
under such conditions.

This plant breeding approach is also worthwhile in that it eliminates the necessity of inoculation. The practice of inoculation,
as performed routinely in the cultivation of legumes, is low in direct cost but may be costly in time and energy if large areas of
land are involved. In addition, although the procedures are simple,
they involve many steps which must all be performed carefully in
order to preclude inoculation failure.

Several conditions that support successful results in the case
of legume inoculation are lacking in the grass-bacteria system.
Important among these is that the Rhizobium-legume system has been
intensively studied for 100 years; inoculation procedures have been
formulated on the basis of a thorough knowledge of the nature and
function of the organisms and the interaction. Further, legumes
are quite selective in their susceptibility to infection, nodulation, and N₂ fixation by strains of Rhizobium. The legume root
therefore provides a strong selective environment for rhizobia.
Plants do not usually appear to select as strongly for associative
bacteria. Having infected the legume root, rhizobia induce nodules
which they occupy in essentially pure culture. This specialized
structure exhibits tissue differentiation, which provides a unique
environment for the survival and functioning of the rhizobia. Vascular elements conduct carbohydrate (energy source) directly to the
bacteria and, in return, receive and transport fixed nitrogen required for plant growth. There is a direct exchange of required nutrients that is excluded from the competitive effects of extraneous
organisms.

Although there may be occupation of cortical tissue by associative bacteria, this may be limited, as exemplified by Azospirillum.
There is increasing evidence that the association is primarily a
rhizoplane phenomenon, both in terms of number of organisms and
primary site of physiological interaction. In this situation, the
transfer of carbon compounds from root to bacteria is indirect and
subject to competitive effects from other rhizosphere microflora.
Likewise, the nitrogen fixed by the bacteria is subject to utilization by the rhizosphere microflora, and organic forms must be min-
eralized before appreciable plant uptake will occur. The conclusion
is that fixation and transfer of nitrogen in the assocative grass-
bacteria system is much less efficient that in the legume-rhizobia
system.

The processes of biological N₂ fixation, like its counterpart
chemical process, requires a great deal of energy. It is quite important, therefore, to consider carefully the possible energy sources
available in biological systems. This is especially true in the grass
system, where many reports of N₂ fixation and plant response are not

easily reconciled with the apparent inefficiency of carbon transfer
to the bacteria. One may ask, "From what sources do the bacteria
receive their carbon and in what amounts?" We have only partial
answers to this question. Evidence indicates that some carbon comes
from the roots in the form of sugars and organic acids released as
exudate or lysate. This is consistent with the observation that
organic acids, such as malate, are favored substrates for Azospirillum in vitro. Another source of carbon compounds is the pectic
polysaccharides extruded from roots by the Golgi apparatus (Wright
and Northcoate, 1974). We do not yet have sufficient data to evaluate fully the capability of Azospirillum or similar organisms to ex-
ploit these energy sources. To this point, inoculation of grasses
has yielded very inconsistent and highly variable results. The high
levels of fixed nitrogen extrapolated from some acetylene-reduction
data are generally not detectable in plant tissue. However, a consistent observation in many trials is that of a significant level of
crop response in the form of an increase in dry matter. It is possible that production of plant growth substances is more important
than N_2 fixation in many cases. Azospirillum produces a complete
range of plant growth hormones in vitro, and it does so in physiologically active amounts. The bacteria induce a great increase in
lateral root and root-hair formation in inoculated grasses. It seems
reasonable to conclude that early formation of an enlarged root system with greatly enhanced capacity for nutrient absorption would
lead naturally to a larger, more vigorous plant. This observation
bears a remarkable similarity to those pertaining to early Russian
attempts to increase crop yields through nitrogen fixation by inoculation with Azotobacter. Some small but consistent yield increases
were ultimately attributed to phytohormone production by the bacteria. They were not a result of N_2 fixation, as was originally
claimed.

In hindsight, the results of inoculation trials with Azospirillum
might have been expected based on the results with Azotobacter. Both
organisms are present in low numbers in agricultural soils. Neither
organism forms an intimate metabolic union with the plant host and
therefore has a limited supply of carbon to provide the large amounts
of energy required for significant N_2 fixation. In both cases, the
organisms are placed in a soil environment not altered in any way to
provide the extra substrate required to support microbial growth and
activity above that already present. Under such conditions, the high
population achieved at inoculation can be expected to die back rapidly to the level of the indigenous organisms - the level inherently
supported by the existing soil environment.

In closing, some comments on research methods will be made for
the possible benefit of those not familiar with problems that may be
encountered. Difficulties in obtaining consistent responses to inoculation were noted earlier. It is obvious that inconsistent responses create doubt about the potential usefulness of the treat-

ments undergoing evaluation. However, in view of the physical and biological complexity of soil-root-interface phenomena, it is not at all surprising that a series of carefully planned and executed experiments might fail to give consistent results. In view of our very superficial knowledge of the mechanisms responsible for plant responses to root-associated bacteria, we should not feel it necessary to immediately find full explanations for every experimental result achieved. The fact that large and highly beneficial responses to treatment occur is fully adequate to justify continued effort to understand these systems, so that they can be used to improve crop yields.

It is important to recognize, however, the importance of using in experiments the best methodology developed to this point. We have already noted that studies with pure cultures provide little useful information about performance of selected strains in the natural environment. In vitro experiments with pure cultures provide needed information about physiological processes, but they have not proved useful as a basis for screening bacterial strains to be further tested as inoculants.

In conducting assays to determine nitrogenase activity, the warnings in the recent literature should be carefully heeded. Excised-root assays may be useful in some instances as rapid, qualitative tests for root-associated organisms capable of fixing nitrogen. However, results of such tests should not be considered indicative of relative abundance or activity of rhizosphere bacteria in the system before removal of the root sample. Any pre-incubation procedure leads to particularly misleading results. Core-sampling procedures are more reliable when properly conducted. However, it is wise to interpret with great care any data on nitrogenase activity acquired with acetylene-reduction assay methods. Such data are highly useful for comparative evaluation of treatment effects and for preliminary screening purposes. They should not be used as a basis for extrapolation or to create figures that purport to show seasonal N₂ fixation rates by associative systems.

It is widely recognized that all screening methods developed thus far have serious limitations. There is no general agreement on whether gnotobiotic culture methods can be used to predict performance under natural conditions. Plant-bacterial interactions that occur in the total absence of competition may not occur when competing organisms are present, if the selected organism does not have a genuine competitive advantage. The question whether such an advantage does in fact exist for Azospirillum or related associative bacteria was addressed earlier. We have never conducted an inoculation test in which Azospirillum could be identified as the dominant bacterium of the rhizosphere. Our enumeration technique may underestimate relative or total numbers of the inoculant strains, but we are reasonably confident of the procedures used. The fact that

substantial plant responses occur, even in experiments in which the
inoculant strain persists only as a minor component in the total
bacterial populations, is one of the many observations still await-
ing adequate explanation.

Several investigators grow plants from sterilized seeds under
gnotobiotic conditions and test interactions by inoculating individual seedings with various bacterial cultures. These methods
normally involve using glass tubes of various sizes for culture
vessels. The root media may be liquid or solid, and solid media
may be agar, sand, soil, or other to suit the purposes at hand. In
most cases, the tubes used are small, and no water or nutrients are
supplied other than those placed in the tubes before seeding. When
tubes 3 to 5 cm in diameter and 20 to 50 cm in length are used, such
tests can be managed easily. Consequently, it is possible to screen
large numbers of combinations. A disadvantage is that there is no
simple method for providing gas exchange, and the plants therefore
grow in a CO_2-depleted environment. We have devised and frequently
used systems in which filtered air is pumped continuously through
the tubes. While this is certainly a desirable procedure, it would
be quite difficult and expensive (but not impossible) to carry out
if large numbers (thousands) of combinations were tested simultaneously. The effort and expense would be acceptable if it were
definitely known that the procedure contains no inherent flaws that
seriously impair its usefulness. At this point, we have, tentatively, an example of a selection made under these conditions which
also has shown evidence that the plant responds similarly to inoculation with the selected bacterial strain under normal culture conditions. This encouraging result suggests that this method may be
satisfactory for screening.

The alternative method is to test plant-bacterial combinations
in the presence of competing organisms; this may be done by growing
the plants in containers or in the field and inoculating by one
method or another with cultures of selected bacterial strains. When
such treatments improve plant yields, no question remains about the
presence of an interaction between the genotypes, provided the test
is conducted properly with adequate controls. For reasons not yet
determined, when such tests are conducted consecutively, the first
experiment frequently shows larger responses to inoculation treatments than do the later tests. It is generally assumed that such
results occur because the inoculant organism, having been cultured
at the experimental site, persists and is present in all plots of
subsequent experiments, including the controls. While this explanation seems reasonable, it has not yet been possible to demonstrate
clearly that this is the case. One aspect of this line of reasoning which limits its plausibility is the implication that, in cases
where large increases in plant yield have resulted from inoculation
of field plots, all plant yields at the experimental site should in
the future reflect the beneficial results indicated by the first ex-

periment. Ths may in fact occur, but it is not easy to demonstrate that this is the case since long-term yield averages, influenced as they are by seasonal variations, are involved.

REFERENCES

J. A. Bassham, Science, 197:630-638 (1977).
J. H. Bouton, R. L. Smith, S. C. Schank, G. A. Burton, M. E. Tyler, R. C. Littell, R. N. Gallaher, and K. H. Quesenberry, Crop Sci., 19:12-16 (1979).
W. J. Brill, Annu. Rev. Microbiol., 29:109-129 (1975).
M. E. Brown, J. Appl. Bacteriol., 35:443-451 (1972).
M. E. Brown, Annu. Rev. Phytopathol., 12:181-197 (1974).
M. E. Brown, J. Appl. Bacteriol., 40:341-348 (1976).
R. H. Brown, in: CO₂ Metabolism and Plant Productivity (C. C. Black, ed.), University Park Press, Baltimore, pp. 311-325 (1976).
J. Dobereiner, Plant Soil., 14:211-217 (1961).
J. Dobereiner, Zentralbl. Bakteriol. Parasitenkd. Infektionskr., Abt. 2, 124:224-230 (1970).
J. Dobereiner and J. M. Day, in: Nitrogen Fixation by Free-Living Microorganisms (W. D. P. Stewart, ed.), Cambridge Univ. Press, Cambridge, pp. 39-56 (1975).
J. Dobereiner and J. M. Day, in: Proc. 1st Intl. Symp. on N₂ Fixation (W. E. Newton and C. J. Nyman, eds.), Washington State Univ. Press, Pullman, pp. 518-538 (1976).
Y. Dommergues and S. V. Krupa, eds., Interactions between Non-Pathogenic Soil Microorganisms and Plants, Elsevier, Amsterdam (1977).
M. H. Gaskins and J. C. Carter, Proc. Soil Crop Sci. Soc. Florida, 35:10-16 (1975).
R. M. Gifford, Aust. J. Plant Physiol., 1:107-117 (1974).
M. G. Hale, L. D. Moore, and G. J. Griffin, in: Interactions between Non-Pathogenic Soil Microorganisms and Plants (Y. R. Dommergues and S. V. Krupa, eds.), Elsevier, Amsterdam, pp. 163-203 (1978).
D. H. Hubbell and M. H. Gaskins, What's New Plant Physiol., 11:17-19 (1980).
E. N. Mishutin, A. N. Naumova, Microbiologia (USSR), 31:543-555 (1962).
J. L. Neal, Jr. and R. I. Larson, Soil Biol. Biochem., 8:151-155 (1976).
C. A. Neyra and J. Dobereiner, Adv. Agron., 29:1-51 (1977).
A. D. Rovira, Bot. Rev., 35:35-37 (1969).
A. D. Rovira, R. C. Foster, and J. K. Martin, in: The Soil-Plant Interface (J. L. Harley and R. Scott Russell, eds.), Academic Press, London, pp. 1-4 (1979).
A. P. Ruschel and R. Ruschel, Proc. Intl. Soc. Sugar Can Technol., 2:1941 (1977).

K. R. Schubert and H. J. Evans, Proc. Natl. Acad. Sci. (U.S.), 73: 1207-1211 (1976).

T. M. Tien, M. H. Gaskins, and D. H. Hubbell, Appl. Environ. Microbiol., 37:1016-1024 (1979).

T. M. Tien, H. G. Diem, M. H. Gaskins, and D. H. Hubbell, Can. J. Microbiol., 27:426-431 (1981).

M. Umali-Garcia, D. H. Hubbell, and M. H. Gaskins, Ecol. Bull. (Stockholm), 26:373-379 (1978).

M. Umaili-Garcia, D. H. Hubbell, M. H. Gaskins, and F. B. Dazzo, Appl. Environ. Microbiol., 39:219-226 (1980).

P. Van Berkum and B. B. Bohlool, Microbiol. Rev., 44:491-517 (1980).

J. F. W. Von Bulow and J. Dobereiner, Proc. Natl. Acad. Sci. (U.S.), 72:2389-2393 (1975).

P. B. Vose and A. P. Ruschel, eds., Associative N_2 Fixation, Vol. 1 and 2, CRC Press, Boca Raton, Florida (1981).

K. Wright and D. H. Northcote, Biochem. J., 139:525-534 (1974).

PHOTOSYNTHESIS AND THE BIOCHEMISTRY

OF NITROGEN FIXATION

Berger C. Mayne

C. F. Kettering Research Laboratory
Yellow Springs, Ohio 45387. U.S.A.

MINIMUM ENERGY REQUIREMENTS FOR NITROGEN FIXATION

Nitrogen fixation is an energetically costly process. This is true whether we are referring to the production of ammonia by the Haber-Bosch process or by biological reactions. In the overall reaction:

$$N_2(gas) + 3H_2O(liquid) \rightarrow 2NH_3(aqueous) + 3/2\ O_2$$

the free energy, ΔG, requirement is approximately 80 kcal per mole of NH_3. As large as this requirement is, it does not represent the total energy requirement for N_2 fixation. In Fig. 1, it can be seen that the energy of $3H_2 + N_2$ is higher than that of $2NH_3$; i.e., the reaction:

$$3H_2(gas) + N_2(gas) \rightarrow 2NH_3(aqueous)$$

is an energy-releasing reaction not an energy-requiring process. The ΔG is -9 kcal/mole NH_3; although the equilibrium is in favor of NH_3, the rate of this reaction at room temperature and atmospheric pressure is not measurable. For the reaction between H_2 and N_2 to proceed, it is necessary to overcome the activation energy barrier between N_2 and NH_3 (Fig. 1). In the Haber-Bosch process, this activation-energy barrier is overcome by the use of a catalyst (generally iron oxide), high temperature, approximately 500°C, and pressures of 12 to 15 atmos (Newton, 1981). These conditions are clearly incompatible with biological systems. Given the high energy requirement, the lability of biological systems, and the extreme sensitivity of nitrogenase to O_2, biological N_2 fixation operates under severe restraints. The chief aim of this paper is to show some of the strategies organisms use to overcome these difficulties in reducing atmospheric N_2.

225

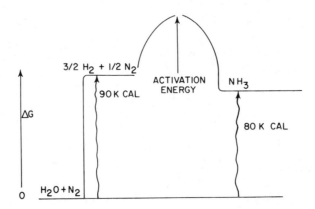

Fig. 1.   Energy diagram for the formation of NH₃ from N₂ and H₂O.
ΔG, free energy change.

The first subject taken up will have to do with obtaining usable
and storable energy, photosynthesis. The second subject will be the
N_2-fixing process.

Good and Bell (1980) state that "Plant productivity is simply a
measure of the total photosynthesis of the plant less any respiration which has occurred during its growth." Thus, in any discussion
of a plant process, it must be remembered that it is ultimately dependent on photosynthesis. Now that we have made a statement on the
overall importance of photosynthesis, it should be pointed out that
a process that takes place in the root is probably only linked to
photosynthesis by sugars formed by photosynthesis and transported to
the root, not by some photoprocess. This makes photosynthesis no
less important but the relation is not as direct.

Since there are any number of good works on photosynthesis at
almost any level of sophistication (Good and Bell, 1980; Govidjee
and Govindjee, 1974; Hatch and Boardman, 1981), this description will
be rather superficial. When necessary because of their relation to
N_2 fixation, elements of the process will be discussed in greater
detail.

The usual formulation for the overall photosynthesis in O_2-
evolving organisms is:

$$N\ h\nu + 6CO_2 + 6H_2O \rightarrow (HCOH)_6 + 6\ O_2$$

where N is the number of quanta necessary to drive the reaction. It
must be remembered that this equation presupposes that the only products of photosynthesis are carbohydrate and O_2. Since this is not
the case, the ratio of O_2 produced to CO_2 taken up will change depending on the degree of oxidation of carbon compounds formed. The

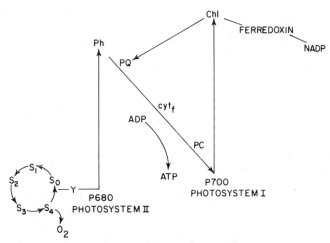

Fig. 2.   Electron transport in photosynthesis. Ph: (Pheophytin);
PQ: plastoquinone; Cyt f: cytochrome f; PC: plastocyanin; Y: electron donor to photosystem II.   See text
for explanation of S_0-$S_4 i$ system.

synthesis of fats and proteins leads to a O_2/CO_2 ratio greater than
one.

The study of photosynthesis may be broken down into two parts:
(a) the reactions that occur between the absorption of light by the
photosynthetic pigments and the oxidation of water to form O_2 and
the reduction of NADP; and (b) the carbon-fixation reactions, from
the reaction of CO_2 with ribulose-1,5-bis-phosphate (RBP) to the
formation of a hexose and the reformation of ribulose-1,5-bis-
phosphate (Bassham and Calvin, 1957).

The electron-transport part of photosynthesis, sometimes referred to as the light reactions, consists of two photoreactions
coupled by an electron transfer chain (Fig. 2) (Hill and Bendall,
1960).

The two photosystems have different dependence on the wavelength of the incident light. Emerson and coworkers (1957) reported
that the drop in the quantum yield of photosynthesis in the red
region of the spectrum could be overcome by adding light of a shorter
wavelength, 670 nm or shorter. Since then it has been shown that the
longer wavelength light is effective in activating photosystem I but
not photosystem II (Wang and Myers, 1976). The problem of the distribution of light energy between the two photosystems is reviewed
by Butler (1978).

In Photosystem II, 4 electrons are transported from water to
the primary electron acceptor, probably pheophytin (Malhis and

Paillotin, 1981). These electrons are then transported through a chain of electron carriers (plastoquinone, cytochrome f, and plastocyanin) to the reaction center of Photosystem I (Avron, 1981) (Fig. 2). The reaction of center pigment of Photosystem I is P700, a special chlorophyll molecule. Electrons from P700 go down another electron-transport chain through ferridoxin(s) to NADP. NADPH is the reductant in the reduction of 1,3-diphosphoglycerate to glyceraldehyde 3-phosphate (Bassham and Calvin, 1957).

The O_2-forming water-oxidizing reactions of photosystem II have a similarity to nitrogenase reactions. In the water-oxidizing reactions, 4 electrons are removed from 2 molecules of water to form 1 molecule of O_2, whereas in the nitrogenase reaction, 6 electrons are taken up to form 2 molecules of NH_3. The water-oxidizing reaction occurs without the formation of any free intermediates. In spite of the lack of free intermediates, it is possible to follow the kinetics of individual reactions. Joliot and Kok measured O_2 evolution following microsecond flashes spaced seconds apart (Kok et al., 1970). They found that little or no O_2 was evolved following flashes 1 and 2 but a maximum amount was evolved following flash 3. The yield per flash then decreased and went through another maximum following flash number 7. By changing the flash spacing, it was possible to study the kinetics of the intermediates, which are usually referred to as "S" states (see Fig. 2). The S_1 states is very stable; therefore, the first maximum occurs following flash number 3 and not flash number 4, as would be expected for a 4 electron requiring process. States S_2 and S_3 decay back to S_1 in the dark, S_4 reacts to give off O_2 and go to S_0. The enzyme involved is thought to be a Mn-containing protein which has not been isolated, at least in an active form (Radmer and Cheniae, 1977).

In addition to the reductant NADPH, ATP is required in the fixation of CO_2 via the Calvin-Benson cycle (the reductive pentose-phosphate cycle). Three ATP and 2 NADPH are required per CO_2 fixed (Bassham and Calvin, 1957). ATP formation is coupled to the photosynthetic electron-transport path between photosystems II and I, probably by the chemiosmotic mechanism of Mitchell (1966).

In isolated chloroplasts, it is possible to demonstrate the occurrence of ATP formation coupled to a cyclic electron flow around photosystem I (cyclic photophosphorylation). Whether this process is of importance in photosynthesis has not been demonstrated. However, it appears that cyclic photophosphorylation may be important in N_2 fixation by cyanobacteria (see below).

CARBON-FIXATION PATH

The basic CO_2 fixation path was worked out by Calvin and coworkers in the 50's, and it survives with minor modifications (Bassham and Calvin, 1957). The basic reactions of photosynthetic

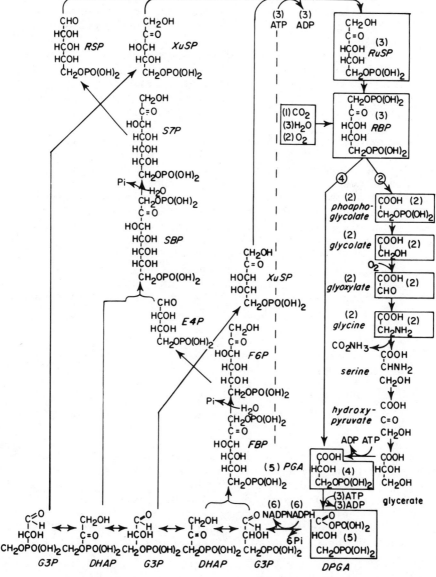

Fig. 3. Photosynthetic carbon metabolism of C_3 plants at the CO_2 compensation point; i.e., the CO_2 concentration at which the rate of CO_2 uptake equals rate of CO_2 evolution. At the compensation point, two molecules of RBP are oxygenated for each one carboxylated. Numbers in parenthesis and circles indicate the relative number of molecules used in the reactions at the CO_2-compensation point. From (Robinson and Walker, 1981) with permission of Academic Press.

carbon fixation via the Benson-Calvin cycle are shown in Fig. 3
(Robinson and Walker, 1981). The first reaction of the Benson-
Calvin cycle is the reaction of CO_2 with ribulose-bisphosphate to
form 2 molecules of 3-phosphoglyceric acid (PGA). This reaction is
catalyzed by the enzyme ribulose-1,5-bisphosphate carboxylase-oxygenase. PGA then reacts with ATP to form 1,3-diphosphoglycerate,
which is reduced by NADPH. The ribulose-1,5-bisphosphate is then
regenerated through a series of reactions as shown in Fig. 3. For
every 6 turns of the cycle, 1 hexose is formed.

The enzyme ribulose-1,5-biphosphate carboxylase-oxygenase also
catalyzes another reaction, the reaction of O_2 with RBP to form 1
molecule of 2-phosphoglycolate and 1 of PGA. This reaction is the
start of a series of reactions which are usually collectively known
as photorespiration (Layzell et al., 1979; Lorimer and Andrews,
1981). These reactions are shown on the lower right-hand side of
Fig. 3. If one examines the reactions, one can see that for every
2 reactions of O_2 with ribulose-1,5-bisphosphate, 1 CO_2 and 3 ATP
are lost. The reactions of photorespiration appear to be wasteful
as far as photosynthesis is concerned, but they may be of importance
in nitrogen metabolism; i.e., in amino acid synthesis. Plants that
have only the Benson-Calvin carbon fixation path are often referred
to as "C_3" plants.

Some plants have developed schemes to minimize the effect of
the oxygenase reaction. One such scheme is that used by the C_4
plants, which was discovered by Kortschak et al. (1965) and Hatch
and Slack (1966). They found that in sugar cane plants, the first
labeled compound formed in photosynthesis was a 4 carbon dicarboxylic
acid, not PGA; this finding led to these plants being designated
"C_4" plants.

It should be pointed out that an important characteristic of C_4
plants (in fact it is one of the diagnostic features used to determine if a plant is C_4) is the occurrence of "Kranz" anatomy, a ring
of chloroplast-containing cells surrounding the vascular bundles in
leaves. These cells are particularly rich in organelles. The division of labor between the bundle sheath cells and mesophyll cells is
at the heart of the C_4 syndrome (Björkman and Berry, 1973; Black,
1971).

In C_4 plants the CO_2 in the substomatal spaces diffuses into
the mesophyll cells where it reacts with phosphoenol pyruvate (PEP)
to form oxaloacetate (OAA). Depending on the species of plant, the
OAA is then either reduced by NADPH to malic acid or aminated to
form aspartic acid (Black, 1971). The C_4 acid formed then diffuses
into the bundle sheath cells, the site of ribulose-1,5-bisphosphate
carboxylase-oxygenase. In the bundle sheath cells, the C_4 acid is
decarboxylated, and the CO_2 released enters the Benson-Calvin cycle.
The mechanism used for the regeneration of PEP depends on the plant
species (Gutierrez et al., 1974).

Since the RBP carboxylase-oxygenase is in bundle sheath cells surrounded by mesophyll cells, any CO_2 released by the oxygenase reaction can only escape by diffusion into the mesophyll cells. In the mesophyll, it will react with PEP and be trapped. The C_4 acid formed can not diffuse from the mesophyll cells into the substomatal space and be lost to the plant.

The CO_2-conserving system operates at the estimated cost of 2 ATP molecules per CO_2 fixed. Ths is in addition to the ATP required for the Benson-Calvin cycle (Edwards et al., 1976).

As a result of these differences, C_4 plants generally have a higher rate of photosynthesis at high light intensity, a more efficient use of water, and a higher temperature optimum (Black, 1971).

Another result of the C_4 pathway is that C_4 plants do not need as much nitrogen as C_3 plants to form the carbon-fixation apparatus, because they are able to fix an equivalent amount of carbon with less RBP carboxylase-oxygenase (Boulton and Brown, 1980), which is the most abundant protein in plants. It is interesting that the only N_2 fixation system associated with C_4 plants is the associative Azospirillum system (Day et al., 1975). Could it be that the reason there are no other N_2-fixing systems related to C_4 plants is that they have developed this strategy for minimizing the need for fixed nitrogen (Brown, 1978)? This strategy is useful for maximizing the production of carbohydrate but is not of much interest for increased protein production.

C_4 photosynthesis occurs in numbers of plant families, both dicotyledonous and monocotyledonous (Downton, 1975). Agronomically the most important C_4 crop plants are Saccharum officinarum (sugar cane), Zea mays (maize), and Sorghum sp. In addition, a large number of important forage plants and noxious weeds are C_4 plants (Black, 1971).

Algae also have a CO_2-concentrating mechanism to either minimize photorespiration or to conserve the CO_2 that would be given off in photorespiration. This mechanism occurs in algae grown under low CO_2 partial pressures; i.e., approximately that of air, 0.03% atmos (Badjer et al., 1980; Lloyd, 1977).

When plants are allowed to photosynthesize in a closed environment until the CO_2 content reaches a photostationary state (i.e., when the rate of evolution of CO_2 by respiration equals the rate of uptake by photosynthesis), the CO_2 concentration reached is called the CO_2-compensation point. In C_3 plants at moderate temperatures and the usual atmospheric O_2 level, the compensation point varies from aproximately 40 to 60 ppm CO_2 in C_3 plants. At higher temperatures and O_2 partial pressures, the CO_2-compensation point of C_3 plants increases. Those plants that have a CO_2 pump have a low CO_2 compensation point (less than 10 ppm CO_2).

Another variant of the carbon fixation path is crassulacean acid metabolism or CAM. In plants with CAM, CO_2 is first fixed into a 4 carbon dicarboxlic acid in a manner similar to that used by C_4 plants; only in this case, it takes place in the dark. During the subsequent light period, the dicarboxylic acid is decarboxylated, and the CO_2 released is refixed by the Benson-Calvin cycle. This enables the plant to close its stomata to conserve water during the heat of the day and still carry on photosynthesis (Radmer and Cheniae, 1977). CAM occurs in a number of plant families. Some of these are Cactaceae, Crassulaceae, Euphorbiaceae, Agavaceae, Bromeliaceae, and Liliaceae (Black and Williams, 1976). Generally, CAM plants are succulent. The most important agricultural CAM plant is pineapple.

Biological N_2 fixation probably appeared only once in evolution. The evidence for this is the similarity of the enzymes in all N_2-fixing organisms. The enzyme nitrogenase is composed of two separate kinds of protein, a MoFe protein and a Fe protein. These proteins have been isolated from a number of procaryotic organisms, from cyanobacteria to Clostridum. All of these enzymes are extremely sensitive to O_2, whether isolated from an O_2-evolving cyanobacterium or an obligate anaerobe such as Clostridium (Yates, 1980).

Nitrogenase proteins are often denoted by an abbreviation consisting of the first letters of the genus and species of the organism: Av for Azotobacter vinelandii followed by 1 for the MoFe protein or 2 from the Fe protein; i.e., Av1 or Av2, respectively, for the two proteins of A. vinelandii (Yates, 1980).

The MoFe proteins have molecular weights ranging from 180,000 in Rhizobium japonicum (Rj1) to 235,000 in Xanthobacter autotrophicus (Xa1). MoFe proteins are made up of 4 subunits, 2 each of 2 kinds and a Mo cofactor (FeMoco) (Yates, 1980). This cofactor is still the subject of research as to its composition and structure (Burgess et al., 1981b). The cofactor is composed of Fe, Mo, and S in a ratio of approximately 7:1:4. The evidence indicates that the isolated cofactor is in much the same form as when it is coupled with the protein. Probably the best evidence for this is the activation of the nitrogenase protein isolated from the mutant A. vinelandii UW45, which does not make the cofactor, by added FeMoco. The reconstructed MoFe protein is then active in the C_2H_2-reduction system (Burgess et al., 1981b).

The Fe protein is smaller and has a molecular weight of from 55,500 in Bacillus polymyxa to 72,600 in Xanthobacter autotrophicus (Yates, 1980). This protein contains a Fe_4S_4 cluster. It is possible to make an active hybrid nitrogenase by combining the FeMo protein of one organism with the Fe protein from a different organism. Although most hybrid nitrogenases are active in catalyzing the usual reactions of nitrogenase, the Fe protein from C. pasteurianum forms a stable, inactive complex with the MoFe protein of A. vinelandii (Burris et al., 1981).

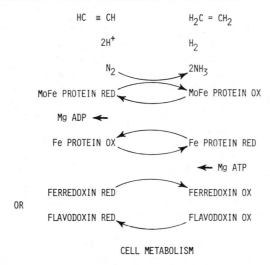

Fig. 4. Electron transport through the nitrogenase system.

TRANSPORT OF ELECTRONS TO THE Fe
PROTEIN OF NITROGENASE

The electron-transport chain through the nitrogenase system is shown in Fig. 4. In the original successful cell-free N_2 fixation experiments, the phosphoroclastic reaction was used to generate the low-potential electrons for N_2 reduction (Carnahan et al., 1960). In aerobic N_2 fixers, the generation of low-potential electrons is thought to be via the tricarboxylic acid cycle (Yates, 1980). Whatever the mechanism for generation of low-potential electrons, it appears that the in vivo electron donor to the Fe protein is either a ferredoxin or a flavodoxin, depending on the species and the Fe supply. Flavodoxin occurs in A. vinelandii when cultured in media containing a small amount of Fe. Most in vitro experiments with nitrogenase are performed with dithionite as the electron donor (Bulen et al., 1965).

Recently, an interesting theory has been advanced for the generation of low-potential electrons for nitrogenase reduction. In this scheme, "the proton motive force across the cytoplasmic membrane, generated by respiration, drives the thermodynamically unfavorable formation of flavodoxin hydroquinone" (Veeger et al., 1981). The evidence for this theory is the inhibition of in vivo nitrogenase activity by ionophores or other uncouplers of phosphorylation that act by dissipating the membrane potential. These might be thought to inhibit N_2 reduction by uncoupling phosphorylation, thereby reducing the availability of ATP, but they have been found to inhibit nitrogenase activity at concentrations which are too low to measurably change the ATP/ADP ratio. Ammonium chloride (0.06 mM) decreased the membrane potential from 106 mV to 95 mV and inhibited nitrogenase activity of Azotobacter vinelandii 60% with little effect on the

ATP/ADP ratio (Veeger et al., 1981). This concentration of ammonium would have no effect on the isolated nitrogenase. Bacteroids
were not affected by ammonium concentrations of up to 10 mM in short-
term experiments, but the ionophore valinomycin inhibited nitrogenase activity at concentrations which had no effect on the ATP/ADP
ratio (Veeger et al., 1981). Experiments with Anabaena variabilis
filaments and isolated heterocysts of the cyanobacterium showed the
same type of inhibition of ionophores (Rowell et al., 1981). The
coupling of the membrane potential to the electron potential in nitrogenase activity is an attractive theory that deserves further research. An alternative to this theory is for the ratio of NADPH/
NADP$^+$ to be large enough for the couple to have a potential sufficiently low to reduce flavodoxin or ferredoxin. The Fe protein is
then reduced by the reduced ferredoxin or flavodoxin.

During reduction of the Fe protein, it reacts with MgATP, reducing the redox potential of the Fe protein. The reduced Fe protein then reacts with the MoFe protein, reducing it. Concurrent
with the reduction of MoFe protein, the MgATP is hydrolyzed to MgADP
(Burris et al., 1981).

In addition to N_2 reduction, nitrogenase catalyzes a number of
other reductions, some of which are: $2H^+$ to H_2; $HCN + 6H^+$ to CH_4 +
NH_3; and $HN_3 + 2H^+$ to $N_2 + NH_3$ (Yates, 1980). The reduction of C_2H_2
to C_2H_4 is of great importance since it is the most widely used
assay for nitrogenase activity. The other reactions of nitrogenase
have been used in the investigation of mechanisms of nitrogenase
activity.

When H_2 reduction is measured with an excess of MoFe protein
over Fe protein, there is a lag in the appearance of H_2 and no lag
in the hydrolysis of MgATP. Burris et al. (1981) convincingly argue
that this and other evidence indicate that the only function of Fe
protein is to reduce the MoFe protein and that "the life time of the
complex between the proteins is no longer than the turnover of the
nitrogenase reductase." The EPR kinetics when compared to the time
course of H_2 evolution also support this explanation (Burris et al.,
1981). Further evidence for the transitory nature of the complex
between the Fe protein and the MoFe protein is the inhibition of the
nitrogenase activity of isolated Azotobacter nitrogenase by Clostridium Fe protein. Clostridum Fe protein forms a stable complex with
Azotobacter MoFe protein which is inactive (Burris et al., 1981).

The number of ATP molecules required per electron transported
is a matter of controversy. Hadfield and Bulen (1969) reported that
4 molecules of ATP were hydrolyzed per electron pair, i.e., 4 ATP
per H_2 molecule formed from hydrogen ions. Recently, this high requirement has been challenged. Mortenson and Upchurch (1981), measuring electron transport, H_2 formation, and ATP hydrolysis as a
function of ADP/ATP ratio, found that as the ratio approached zero,

4 molecules of ATP were hydrolyzed per electron pair, whereas at
a ratio of approximately 0.5, the requirement was 2 molecules of
ATP per electron pair. The ATP requirement then increased as the
ADP/ATP ratio increased. Hadfield and Bulen (1969) used an ATP-
generating system which probably maintained the ADP/ATP ratio at
a value less than that required for the minimum ATP requirement.

Redox studies of the FeMo protein show that the protein will
take up to 6 electrons (Watt et al., 1980). This fits very well
with the number required to reduce 1 N_2, but if as suggested, 1 H_2
is reduced per each fixed N_2, it does not fit. The in the reduction of N_2, it is bound to the FeMo protein; where on the protein
is not known with any degree of certainty, but it is presumed to be
the FeMo cofactor. The FeMo protein-N_2 complex is reduced by the
reduced Fe protein, and 2 NH_3 are released. This is an over-
simplified statement of the process, but it gives the overall reaction leading to NH_3 formation (see D_2H_2 exchange reaction).

The so called exchange reaction, the formation of HD from D_2
by nitrogenase (Burgess et al., 1981a), has been useful in the investigation of probable intermediates of N_2 fixation. This reaction has been shown to have two components, a minor one independent
of N_2 and a major one dependent on the N_2. Both reactions require
active electron transport to the nitrogen system. It was also shown
that the exchange reaction is not an exchange with H_2O. All the
evidence is compatible with the reaction being the removal of 2H
from an intermediate in a reaction of the following type:

$$\underset{N}{\overset{H}{\diagdown}} = \underset{N}{\overset{H}{\diagup}} + D_2 \rightarrow N \equiv N + 2HD$$

The evidence is consistent with the hypothesis that a diazene-type
intermediate occurs in N_2 fixation, and this mechanism explains the
inhibition of N_2 fixation by H_2. Thorneley and coworkers (1978)
reported the formation of an enzyme-bound dinitrogen hydride (N_2H_2)
intermediate. The kinetics of the formation of the dinitrogen hydride were compatible with its being an intermediate in the reduction of N_2 to NH_3. There has been considerable research on the
mechanism of the reduction of N_2, but most of the evidence for any
one scheme is still equivocal.

An important and interesting group of N_2 fixers are the N_2-
fixing cyanobacteria. In these organisms, there is the possibility
for a closer coupling between the energy-accumulating photosystems
and the energy-requiring system, N_2 fixation. This also required
the development of a mechanism for the protection of the nitrogenase
from the O_2 produced by photosynthesis (Gallon, 1981). The most important N_2-fixing cyanobacteria, at least as far as the number of
species, are the filamentous forms with heterocysts. In these or-

ganisms, at least when grown aerobically in the absence of combined
nitrogen, the nitrogenase activity is limited to specialized thick-
walled cells called heterocysts. Heterocysts lack ribulose-1,5-
bisphosphate carboxylase and Photosystem II; i.e., the O_2-evolving
system appears to be absent. The absence of the Photosystem II helps
solve the problem of the protection of nitrogenase from the O_2
evolved in photosynthesis.

Since heterocysts lack Photosystem II, Photosystem I can not
continue to reduce the nitrogenase Fe protein in the absence of an
electron donor other than water. There has been speculation that
some intermediate of carbon metabolism might act as a donor to Photosystem I, leading to the reduction of ferredoxin, which acts as
the reductant of the Fe protein. I do not think that there is any
evidence for this occurring to the extent necessary to support
steady-state N_2 fixation at the rates observed. Presently, it is
thought that the low-potential electrons for the reduction of the
nitrogenase in the heterocysts are generated by the metabolism of
carbohydrates transported from the vegetative cells to the heterocysts (Wolk, 1980).

It is clear that N_2 fixation in heterocystous
greatly increased in the light. Therefore, photosynthetic processes
must be important in N_2 fixation. If, as appears likely, there is
no photo-driven noncyclic electron transport of electrons in the
heterocysts, the photosynthetic electron transport coupled to phosphorylation is probably cyclic; i.e., electrons from near the top
of Photosystem I return to the chain between photosystems, probably
at the level of plastoquinone. Cyclic photophosphorylation has been
well documented in reactions of isolated chloroplasts, but the extent of its occurrence in vivo is unknown. The occurrence of ATP-
requiring reactions after the inhibition of Photosystem II with DCMU
(Diuron) is usually taken as evidence for cyclic photophosphorylation.

Fay (1970) measured the nitrogenase activity of Anabaena cylin-
drica as a function of wavelength under light-limiting conditions,
and he plotted the results as rate of C_2H_2 reduction per incident
energy vs wavelength of the monochromatic light used. The action
spectrum obtained showed that the reaction is a Photosystem I reaction. The decrease in the rate of the reduction of C_2H_2 per incident light energy occurred at longer wavelengths than the decrease
in photosynthesis which requires both Photosystems I and II. In
cyanobacteria and other phycobiliprotein-containing organisms, the
efficiency in the use of light by Photosystem II and therefore photosynthesis decreases in that part of the spectrum where chlorophyll
is the only pigment absorbing light (Wang et al., 1977). Tyagi et
al. (1981) measured the action spectrum of C_2H_2 reduction by Anabaena
azolla isolated from Azolla caroliniana. This also indicated that
N_2 fixation was a Photosystem I process.   These results indicate

that the electrons for the reduction of the Fe protein come from carbon metabolism and a greater part of the ATP is supplied by cyclic photophosphorylation. Action spectra also indicated that light absorbed by phycobiliproteins was at least 70% as effective as light absorbed by chlorophyll (Fay, 1970). The action spectrum of the nitrogenase activity of nonheterocystous cyanobacteria should be obtained to determine if it is possible that nitrogenase may be reduced directly by the photosynthetic electron-transport chain in these organisms.

Previously, I have said that the free-energy requirement for N_2 fixation (i.e., the formation of NH_3 from N_2 and H_2O) was approximately 80 kcal per mole NH_3, but at the same time I indicated the need for energy to overcome an unspecified but large activation-energy barrier. Since then, I have indicated the need for energy in addition to that required for generation of the low-potential reductant by the metabolism of the organism; i.e., 2 to 4 ATP molecules per electron pair depending on which estimate of the ATP requirement is correct.

If one assumes that the usually quoted ATP requirement is correct (i.e., 4 ATP molecules per electron pair), the requirement is 6 per NH_3 formed, and since the formation of at least one more H_2 per NH_3 seems to be obligatory, this means 2 more or a total of 8 ATP per NH_3; however, this is still much too low an estimate of the energy requirement for N_2 fixation.

If the activation of NADH by membrane potential is important in the reduction of the Fe protein, how many molecules of ATP would this be equivalent to? We may estimate a requirement of approximately 1 ATP per electron. An alternative to membrane activation is the maintenance of a high NADPH/NADP ratio; this would also require additional energy. In addition to the directly assignable energy requirement for N_2 fixation, there are a number of overhead costs, such as the maintenance of the nodule.

Before proceeding any further in the discussion of the energy cost of N_2 fixation, I should discuss the obligatory H_2 evolution from hydrogen ions and the uptake of the H_2 by hydrogenase to reclaim some of the energy that would be wasted. Recall that one of the reactions of nitrogenase is the reduction of hydrogen ions to molecular hydrogen. This reaction appears to be a required part of the reduction of nitrogen. The loss of H_2 in N_2 fixation represents a loss of energy equal to from a third to a half of that used in the fixation of NH_3. Some organisms have developed a mechanism for metabolizing H_2 and reclaiming some of the energy that would be lost. Evans et al. (1981) review this subject.

The presence of an uptake hydrogenase to metabolize the H_2 released increased the total N_2 fixed up to 49% in soybeans inoculated

with the desired strains and cultured in Leonard jars. The R.
japonicum strains lacking the H_2-uptake systems were derived from
the uptake hydrogenase-positive strain. In another experiment, 5
strains containing hydrogenase were compared with 5 lacking the uptake system. In this experiment, the increase in total nitrogen was
26%.

Phillips (1980) collected the reported values for the efficiency of N_2 fixation (g C/g N) by legumes. These values ranged
from 20 to 0.3 g C/g N, with a large number between 6 and 8 g C/g N
or 1.2 to 1.5 mole glucose per mole NH_3. The usual method for determining the cost of N_2 fixation in nodules is to measure the respiration of the nodules and the nitrogenase activity (usually C_2H_2
reduction), and from these measurements, after making a number of
assumptions, to estimate the cost of N_2 fixation.

Mahon (1977a, 1977b, 1979) measured respiration (CO_2 evolution)
and C_2H_2 reduction while subjecting plants to treatments that affected the rate of respiration. The CO_2 evolution was then plotted
against C_2H_2 reduction. The slope of the straight-line fit of the
data had the dimensions of CO_2 evolved per C_2H_2 reduced, or the cost
of the operation of nitrogenase. For treatments that affected the
amount of carbon transported to the nodule, light/dark, diurnal
change, age of plant, and partial defoliation, the cost of C_2H_2 reduction was from 0.23 to 0.43 mg CO_2 per μmole C_2H_2 reduced or 1.31
to 2.44 mole glucose per mole NH_3. When he treated plants with nitrate or ammonium, comparison of C_2H_2 reduction with CO_2 evolution
gave an average value of 17 g carbohydrate per g N_2 or 1.32 mole
glucose per mole NH_3.

Layzell et al. (1979) in a detailed analysis of carbon and nitrogen metabolism of cowpea, Vigna unguiculata, inoculated with
Rhizobium CB756, found that 1.54 mg C and "negligible H were evolved"
per mg nitrogen fixed, or 0.299 mole glucose per mole NH_3. In the
lupin, Lupinus albus, 3.84 mg C was metabolized and 0.22 mg H_2
evolved per mg N_2 fixed, or 0.71 mole glucose per mole NH_3. The results of their investigation are summarized in Fig. 5. In addition
to the measurement of the total N_2 fixed and gases evolved (CO_2 and
H_2), they measured the increase in nitrogen and carbon content of
the nodules and the transport of carbon and nitrogen compounds to
and from the root through the xylem and phloem. Except for CO_2 and
H_2 evolution, values were determined by periodic sampling of the
population. The major qualitative difference between the two symbioses is the lack of H_2 evolution by the cowpea symbiosis and the
difference in the nitrogen compounds exported from the nodule. In
cowpea, ureides (allontoxin and allontoxic acid) were "the principal
compounds exporting fixed N," whereas in lupin, "asparagine was the
major solute of xylem sap" (Layzell et al., 1979).

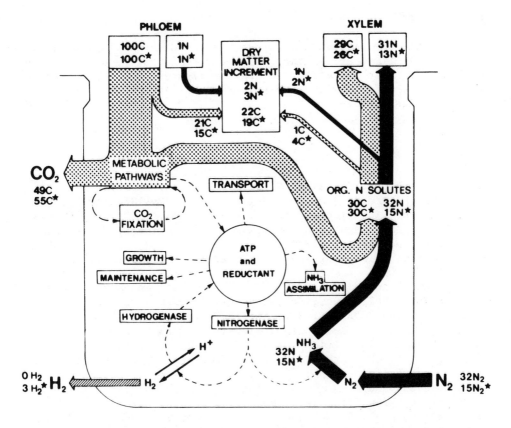

Fig. 5. Carbon and nitrogen budget of cowpea (<u>Vigna unquiculata</u>) –
<u>Rhizobium</u> CB756 symbiosis and of the white lupin (<u>Lupinus</u>
<u>albus</u>) – <u>Rhizobium</u> 425 symbiosis. All values are expressed
as mg per 100 mg of C translocated as carbohydrate. From
Layzell et al. (1979) with permission of the publisher.

        The relation of photosynthesis to N_2 fixation has been investi-
gated by a number of workers. Havelka and Hardy (1975) have shown
that when N_2-fixing soybean plants are grown in an atmosphere enriched with CO_2 (from 800–1200 ppm CO_2) starting at an age of 38
days, there was an approximately five-fold increase in total N_2 fixed
compared to the control. Williams et al. (1981), on the other hand,
showed that in the first 22 days after planting, the addition of CO_2
(1000 ppm) had no effect on soybean plants grown on N_2 or 1 mM NH_4NO_3,
but the CO_2 increased the rate of growth of plants grown on 8 mM
NH_4NO_3. These results indicate that young N_2-fixing soybean plants
are nitrogen-limited, whereas older plants are limited by photosynthesis.

ACKNOWLEDGMENTS

I thank the following coworkers at the C. F. Kettering Research Laboratory: Darrell Fleischman and Becky Poole for reading the manuscript; Marsha Tootle for preparing the figures, and Debbie Patten for typing the manuscript. Contribution No. 761 from the C. F. Kettering Research Laboratory.

REFERENCES

S. L. Albrecht, R. J. Maier, F. J. Hanus, S. A. Russell, D. W. Emerich, and H. J. Evans, Science, 203:1235-1257 (1979).
M. Avron, in: The Biochemistry of Plants, Vol. 8, Photosynthesis (M. D. Hatch and N. K. Boardman, eds.), Academic Press, New York, pp. 164-191 (1981).
M. R. Badger, A. Kaplan, and J. A. Berry, Plant Physiol., 66:407-413 (1980).
J. A. Bassham and M. Calvin, The Path of Carbon in Photosynthesis, Prentice Hall, Inc., Englewood Cliffs, New Jersey (1957).
O. Björkman and J. Berry, Sci. Am., 229(4):80-93 (1973).
C. C. Black, Advan. Ecol. Reg., 7:87-113 (1971).
C. C. Black and S. Williams, in: CO_2 Metabolism and Plant Productivity (R. H. Burris and C. C. Black, eds.), University Park Press, Baltimore, pp. 407-408 (1976).
J. K. Boulton and R. H. Brown, Plant Physiol., 66:97-100 (1980).
R. H. Brown, Crop Sci., 18:93-98 (1978).
W. A. Bulen, R. C. Burns, and J. R. LeComte, Proc. Nat. Acad. Sci. (U.S.), 53:532-539 (1965).
B. K. Burgess, S. Wherland, W. E. Newton, and E. W. Stiefel, Biochemistry, 20:5140-5146 (1981a).
B. K. Burgess, S.-S. Yang, C.-B. You, J.-G. Li, G. D. Friesen, W.-H. Pan, E. I. Stiefel, and W. E. Newton, Current Perspectives in Nitrogen Fixation (A. H. Gibson and W. E. Newton, eds.), Australian Academy of Sciences, Canberra, pp. 71-74 (1981b).
R. H. Burris, D. J. Arp, R. V. Hageman, J. P. Houchins, W. J. Sweets, and M. Tso, in: Current Perspectives in Nitrogen Fixation (A. H. Gibson and W. E. Newton, eds.), Australian Academy of Science, Canberra, pp. 52-66 (1981).
W. L. Butler, Annu. Rev. Plant Physiol., 29:345-378 (1978).
J. E. Carnahan, L. E. Mortenson, H. F. Mower, and J. E. Castle, Biochim. Biophys. Acta, 44:520-535 (1960).
J. M. Day, M. C. P. Neves, and J. Döbereiner, Soil Biol. Biochem., 7:107-112 (1975).
W. J. G. Downton, Photosynthetica, 9:96-105 (1975).
G. E. Edwards, S. C. Huber, S. B. Ku, C. K. M. Rathnam, M. Gutierrez, and B. C. Mayne, in: CO_2 Metabolism and Plant Productivity (R. H. Burris and C. C. Black, eds.), University Park Press, Baltimore, pp. 83-112 (1976).

R. Emerson, R. Chalmers, and C. Cederstand, Proc. Nat. Acad. Sci.
    (U.S.), 43:133-143 (1957).
H. J. Evans, K. Purohit, M. A. Cantrell, G. Eisbrenner, S. A.
    Russell, F. J. Hanus, and J. E. Lepo, in: Current Perspectives
    in Nitrogen Fixation (A. H. Gibson and W. E. Newton, eds.),
    Australian Academy of Science, Canberra, pp. 84-96 (1981).
P. Fay, Biochim. Biophys. Acta, 216:353-356 (1970).
J. R. Gallon, T.I.B.S., 6:19-23 (1981).
N. E. Good and D. H. Bell, Biology of Crop Productivity (P. S.
    Carlson, ed.), Academic Press, New York, pp. 3-51 (1980).
B. Govindjee and R. Govindjee, Sci. Am., 231(6):68-80 (1974).
M. Gutierrez, V. E. Gracen, and G. E. Edwards, Planta, 119:279-300
    (1974).
K. L. Hadfield and W. A. Bulen, Biochemistry, 8:5103-5108 (1969).
M. D. Hatch and N. K. Boardman, eds., The Biochemistry of Plants,
    Vol. 8, Photosynthesis, Academic Press, New York (1981).
M. D. Hatch and C. R. Slack, Biochem. J., 101:103-111 (1966).
U. D. Havelka and R. W. F. Hardy, in: Proceedings of the 1st Inter-
    national Symposium on Nitrogen Fixation (W. E. Newton and C. J.
    Nyman, eds.), Washington State University Press, Pullman, Wash-
    ington, pp. 456-475 (1975).
R. Hill and F. Bendall, Nature (London), 186:136-137 (1960).
B. Kok, B. Forbush, and M. McGloin, Photochem. Photobiol., 8:457-
    475 (1970).
H. P. Kortschak, C. F. Hardt, and G. O. Burr, Plant Physiol., 40:
    209-213 (1965).
D. B. Layzell, R. M. Rainbird, C. A. Atkins, and J. S. Pate, Plant
    Physiol., 64:888-891 (1979).
N. D. H. Lloyd, D. T. Canvin, and D. A. Culver, Plant Physiol., 59:
    936-940 (1977).
G. H. Lorimer and J. J. Andrews, in: Biochemistry of Plants, Vol.
    8, Photosynthesis (M. D. Hatch and N. K. Boardman, eds.),
    Academic Press, New York, pp. 329-374 (1981).
J. D. Mahon, Plant Physiol., 60:812-816 (1977a).
J. D. Mahon, Plant Physiol., 60:817-821 (1977b).
J. D. Mahon, Plant Physiol., 63:892-897 (1979).
P. Malhis and G. Paillotin, in: The Biochemistry of Plants, Vol.
    8, Photosynthesis (M. D. Hatch and N. K. Boardman, eds.),
    Academic Press, New York, pp. 98-161 (1981).
P. Mitchell, Biol. Rev., 41:445-502 (1966).
L. E. Mortenson and R. G. Upchurch, in: Current Perspectives in
    Nitrogen Fixation (A. H. Gibson and W. E. Newton, eds.),
    Australian Academy of Science, Canberra, pp. 75-83 (1981).
W. D. Newton, in: Encyclopedia of Chemical Technology, 3rd Ed.,
    Vol. 15 (M. Grayson and D. Eckroth, eds.), Wiley-Interscience,
    New York, pp. 942-962 (1981).
D. A. Phillips, Annu. Rev. Plant Physiol., 31:29-49 (1980).
R. Radmer and G. M. Cheniae, in: Topics in Photosynthesis, Vol. 2
    (J. Barber, ed.), Elsevier Scientific Publishing Co., Amsterdam,
    pp. 303-349 (1977).

S. P. Robinson and D. A. Walker, in: The Biochemistry of Plants,
    Vol. 8, Photosynthesis (M. D. Hatch and N. K. Boardman, eds.),
    Academic Press, New York, pp. 193-236 (1981).

P. Rowell, R. H. Reed, M. J. Halukesford, A. Ernst, J. Diez, and
    W. D. P. Stewart, in: Current Perspectives in Nitrogen Fixa-
    tion (A. H. Gibson and W. E. Newton, eds.), Australian Academy
    of Science, Canberra, pp. 186-189 (1981).

R. N. F. Thorneley, R. R. Eady, and D. J. Lowe, Nature (London),
    273:557-579 (1978).

I. P. Ting, in: CO_2 Metabolism and Plant Productivity (R. H. Burris
    and C. C. Black, eds.), University Park Press, Baltimore, pp.
    251-268 (1976).

V. V. S. Tyagi, T. B. Ray, B. C. Mayne, and G. A. Peters, Plant
    Physiol., 69:1479-1484 (1981).

V. Veeger, H. Haaker, and C. Laane, in: Current Perspectives in
    Nitrogen Fixation (A. H. Gibson and W. E. Newton, eds.),
    Australian Academy of Science, Canberra, pp. 101-104 (1981).

R. T. Wang and J. Myers, Photochem. Photobiol., 23:411-414 (1976).

R. T. Wang, C. L. R. Stevens, and J. Myers, Photochem. Photobiol.,
    25:103-108 (1977).

G. D. Watt, A. Burns, S. Lough, and D. C. Tennent, Biochemistry,
    21:4926-4932 (1980).

L. E. Williams, T. M. DeJong, and D. A. Phillips, Plant Physiol.,
    68:1206-1209 (1981).

C. P. Wolk, in: Nitrogen Fixation, Vol. 2, Symbiotic Associations
    and Cyanobacteria (W. E. Newton and W. H. Orme-Johnson, eds.),
    University Park Press, Baltimore, pp. 279-292 (1980).

M. G. Yates, in: The Biochemistry of Plants, Vol. 5, Amino Acids
    and Derivatives (B. J. Miflin, ed.), Academic Press, New York,
    pp. 1-64 (1980).

INDEX

243

Date Due